Workbook for
TECHNOLOGY OF
MACHINE TOOLS

SEVENTH EDITION

Workbook for
TECHNOLOGY OF
MACHINE TOOLS

SEVENTH EDITION

Steve F. Krar
Arthur R. Gill
Peter Smid

WORKBOOK FOR TECHNOLOGY OF MACHINE TOOLS, SEVENTH EDITION
Steve F. Krar, Arthur R. Gill, and Peter Smid

Published by McGraw-Hill, a business unit of The McGraw-Hill Companies, Inc., 1221 Avenue of the Americas, New York, NY 10020.
Copyright © 2011 by The McGraw-Hill Companies, Inc.
All rights reserved. Previous editions © 2005, 1997, 1990, 1984, 1976, and 1969.

Printed in the United States of America.

3 4 5 6 7 8 9 QVR/QVR 19 18 17 16 15 14 13

ISBN   978-0-07-738988-8
MHID   0-07-738988-3

www.mhhe.com

TMT 7TH WORKBOOK TABLE OF CONTENTS

TMT 7TH WORKBOOK TABLE OF CONTENTS

TO THE INSTRUCTOR

The world of manufacturing has changed dramatically in the last 50 years and this change has been very noticeable in the last 5 to 10 years. The catchword in manufacturing is *Lean* because it involves the identification and elimination of waste and the continuous improvement of all operations involved in manufacturing.

Global competition has created a need for industry to produce high-quality products quickly and at less cost. The implementation of Lean production systems has saved companies millions of dollars over the last 20 years. The waste in how people, materials, and manufacturing processes are used can amount to 35 percent or more of a company's revenue.

Technology of Machine Tools, Seventh Edition, covers the basics of six Lean tools that students should be made aware of before they enter the world of work in any field. Questions on this important aspect of today's manufacturing are included in the text and workbook.

This workbook is designed to be used with the text *Technology of Machine Tools,* Seventh Edition. It contains a variety of tests and quizzes covering subjects discussed in the text. There are five general types of questions: multiple choice, true or false, completion, identification, and matching. Each test is designed to review a topic thoroughly, to stimulate students' thinking, and to increase their knowledge in each area.

In the interests of fairness to the student and for ease of grading, it is the intention of the authors that each question be of equal value. With this in mind, the value of each question and the time allotted for each test has been left to the discretion of the instructor. Generally, a test or quiz should take the student about three times longer than it would take an instructor to answer the same test. The number under the score line for each test represents the relative value of each test for grading purposes.

The tests are arranged so that all the answers are placed on the right-hand side of the page. This format allows an easier method of checking student answers and a more uniform method of recording answers.

An instructor's answer key for this book is available from the publisher. The wording of student answers for some questions may vary somewhat from the authors', and therefore the instructor should use discretion in deciding whether a student answer is satisfactory and should mark it accordingly.

These tests should provide a good indication to the instructor as to the areas of difficulty encountered by each student. With this information, the instructor can then clarify any points that have been misunderstood.

The projects in this workbook are designed to provide students with a variety of basic learning experiences using common machine and hand tools. The operational procedures for each project are designed to fit a general shop used for training purposes. If the suggested procedures do not suit the equipment in your shop, please feel free to change the operational procedures accordingly. In some projects, you may feel that some parts of the operations are beyond the development stage or capabilities of the student at that time. Do not hesitate to do a certain portion of the project and have the student complete the remainder when the necessary knowledge or skills have been attained.

The ways in which instructors use a text as a teaching and learning tool are bound to vary due to their individual differences and background experiences. Since no one method of teaching suits everyone, instructors ought to use the method of instruction that works best for them. To be effective, however, an instructor should use as many teaching aids as possible, and actively involve students in the learning process.

Steve Krar
Arthur Gill
Peter Smid

A Guide to the Use of This Workbook

The questions in this book are for use with the text *Technology of Machine Tools,* Seventh Edition. There are five general types of questions: multiple choice, completion, true or false, identification, and matching. In order to assist you in answering them correctly, sample questions and answers are provided. You should study these sample questions carefully in order to understand the procedure required for answering each type.

TYPE 1—MULTIPLE CHOICE

There are four choices for each question. Select the most appropriate answer for each question and circle the letter in the right-hand column that indicates your choice.

Example   *Answer*

1. The most common hammer used by a machinist is a   1.   A   B   C   (D)
   (A) soft-faced hammer   (C) claw hammer
   (B) toolmaker's hammer   (D) ball-peen hammer

   The correct answer is *ball-peen hammer;* therefore, the letter D should be circled in the right-hand column as shown.

TYPE 2—SENTENCE COMPLETION

These questions contain blank spaces that must be filled in with the proper word or words to make each sentence complete and true. The answer is recorded in the blank space provided at the right-hand side of the page.

Example   *Answer*

2. A box-end wrench has ___?___ precisely cut notches around the inside face.   2.   ____12____

   The correct answer is 12 and should be inserted in the space provided at the right-hand column as shown.

TYPE 3—TRUE OR FALSE

In this type of question, a statement is made. You must determine whether the statement is true or false and circle the appropriate letter (T for true and F for false) in the right-hand column.

Example   *Answer*

3. Revolving work may be measured safely.   3.   T   (F)

This statement is *false* so the letter F should be circled as shown.

TYPE 4—IDENTIFICATION

This type of test contains photographs or labeled drawings, the parts of which must be identified by placing the appropriate name or letter in the right-hand column.

Example

4. Record the parts of the drill indicated by the letters on the illustration.

Answer

4. A _____ body _____

B _____ shank _____

TYPE 5—MATCHING

These tests contain photographs or drawings, each identified by a letter on the left-hand side of the sheet. In the right-hand column there is sufficient information to identify the photograph. You must match the *letter* on the photograph with the proper *information.*

Example

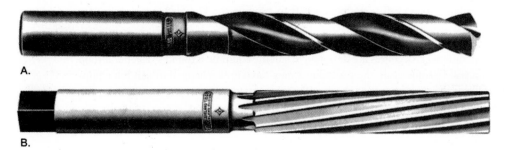

A.

B.

Answers

5. used to finish a hole _____ B _____

6. used to produce a hole _____ A _____

Figure A shows a drill used to produce a hole. The letter A is then matched with the number 6. The reamer (B) is used to finish a hole; therefore, the letter B is matched with the number 5.

TO THE STUDENT

The changing technological world of today is very different than what it was as little as 5 or 10 years ago. Advanced technology and global competition are the driving forces of this never-ending change. In order to succeed in the 21st century, the manufacturing operations must be Lean, global, modern, technology-focused, highly flexible, quality-driven, and cost competitive, with a well-trained workforce. Lean is a concept that commits a company to a program that eliminates waste, simplifies processes, improves quality, and speeds up production.

Some manufacturers today are playing catch-up with progressive companies who saw the need for change many years ago (some over 50 years) and streamlined their operations with Lean tools and advanced machine processes and information technology. These companies have benefited by increasing their productivity, producing high-quality goods and providing on-time delivery.

Quick and easy kaizen is a simple and effective way to encourage people to come up with small ideas to change their jobs for the better. This inspires the employees and encourages them to experiment to experience the fun and satisfaction of coming up with a new idea that can solve a problem, to implement a solution and write up the idea to display, and to share with others. It will challenge employees to turn their own ideas into success for the company and for themselves. The end result of this exercise is with happier people, a big improvement in people's attitudes about themselves, productivity, quality; and safety will improve; costs will go down; and the customers will be satisfied.

This workbook is designed to be used along with the text *Technology of Machine Tools,* Seventh Edition. It contains tests and quizzes on each phase of your development in the machine tool trade. The tests are designed to test your knowledge about topics and provide the background for developing technical (practical) skills. In places, references are made to various sections or units in the text. They are designed to direct you to the proper area in the book to refresh your memory or to provide the knowledge necessary for producing quality work.

The projects included in this workbook are designed to give basic training in the use of various machine and hand tools. The project operational procedures are designed to suit the equipment found in a general shop that is used for training purposes. In some cases the procedures may have to be changed to suit the equipment found in your shop. If this is the case, be sure to check with the instructor on the best procedures to use. The instructor can suggest equipment that can be used to perform a specific operation.

GENERAL SUGGESTIONS

The following suggestions are offered to assist each person to develop as fully as possible to eventually become a machinist or tool and diemaker.

1. Work Quality and Quantity
- Develop pride in producing the best quality work possible.
  - -In each stage of your training and development, the quality of your work should improve.
- Always try to produce as much work as is reasonably possible. *Wasted Time is Lost Time.*
  - -Never develop the attitude that you will do only enough to get by. This attitude can hinder you throughout your life.
  - -In industry, quality and quantity go hand in hand; without both, a company could soon be out of business.

2. Dimensional Accuracy
- Learn how to use measuring tools and machine tools graduated micrometer collars as quickly as possible.
  - -A little practice on the use of these tools can make it possible to produce work to within $\pm$ .001-in. accuracy early in your training program.
  - -Always take a light trial cut about .250 in. long from the part to check the accuracy of the tool setting.
  - -*Always measure before machining a diameter or surface.*

3. Work Attitudes
- Learn about the job requirements of the career in which you hope to specialize.
    - Good work habits are a prime requirement for successful workers.
    - Good work habits are as important to a technician as are technical skills.

4. Conscientious Attendance
- It is important for success in school and industry to develop the habit of being prompt and reliable at all times.
    - Regular attendance will give you the best chance for success.

5. Relationship with Others
- The ability to work as part of a successful team is important in school, in industry, and throughout your life.
    - Understand the importance of a team effort.
    - Since each person is different, learn to respect the feelings and needs of each individual.
    - Most new workers lose jobs because they cannot get along with fellow workers or the supervisor.

6. Job Planning
- Carefully review the two sections on job planning that follow to learn the procedures for machining round or flat parts.
    - Prepare yourself for a successful future by starting to plan operational sequences early in your training.
    - In the early stages of your development, these operational sequences will be provided in detail.
    - In later stages, especially in industry, you will be given a technical drawing and be expected to plan the operational sequences for each part.

7. New Technology
- Keep pace with new technology if you expect to be successful in your career.
    - Each person must develop a desire for learning new technologies and the skills required to apply them.
    - Read technical and trade publications to learn about developments in the trade or profession you may be planning to enter.
- Join technical organizations such as the Society of Manufacturing Engineers (SME) that have continual updating programs on new technologies.

It is the sincere wish of the authors that you will find the text *Technology of Machine Tools,* Seventh Edition and this *Workbook* helpful in becoming a machinist or tool and diemaker. If you should have any suggestions on how the text or workbook can be changed to make them more useful to you, please send your suggestions to the publisher.

DIMENSIONING STANDARDS (PRACTICES)

Ever since NC (numerical control) was introduced in the late 1950s, there has been a trend to use decimals to standardize dimensioning throughout the world. Since the early 1970s, ANSI (American National Standards Institute) has recommended that the decimal inch system (in.) be used in .02 in. increments to replace fractional inch dimensions. However, common fractions may still be used when producing holes that require the use of drills stocked in fractional sizes and for the sizes of standard screw threads.

The drawings in the *Workbook for Technology of Machine Tools,* Seventh edition, which do not use fractional dimensioning, introduce a few of the modern drawing practices of our ever-changing technological world. These include decimal dimensioning and limit dimensioning, along with some drawing symbols that have become widely accepted throughout the world and have helped convert complex drawing standards into understandable instruction units.

The new standards that apply to the drawings in this book involve manufacturing methods, tolerances, and the use of symbols.

Manufacturing Methods

The drawing only defines a part but does not specify the manufacturing method used to perform operations to produce a part. For example, only the diameter of a hole is given without indicating whether it should be drilled, reamed, bored, or produced by any other method. However, if a dimension is critical, its tolerance or limits are provided on the drawing so that the craftsperson will use the method to produce the part to the accuracy required.

Dimensioning Standards

Each dimension on a drawing should have a tolerance that defines the accuracy of a specific operation or part. It is accepted machine trade practice that the tolerance on dimensions is usually a + or − unit of the last digit; for example:

- .22 would indicate a tolerance of ± .010 in.
- .225 would indicate a tolerance of ± .001 in.
- .2256 would indicate a tolerance of ± .0001 in.

Inch Dimensions

- All fractional sizes are stated to two decimal places—for example, 3/4 in. is stated as .75 in., which indicates that it is not a critical size.
- Whole dimensions are shown with a minimum of two zeros to the right of the decimal point; for example, 7.00 in., not 7 in.
- No zero is used to the left of the decimal for any value of less than one inch; for example, .72 in., not 0.72 in.; or .725 in., not 0.725 in.
- Sizes that are critical dimensions are shown with three or four decimal places. Where necessary, the tolerance or limit dimensions are included.

Metric Dimensions

- A zero must be used to the left of the decimal for all sizes less than one millimeter; for example, 0.15 mm, not .15 mm.
- Where the dimension is a whole number, no decimal point or zero follows the number; for example, 2 mm, not 2.0 mm.
- Where the dimension is over the whole number by a decimal fraction, the last digit to the right of the decimal point is not followed by a zero; for example, 4.5 mm, not 4.50 mm.

Symbols

A few of the common symbols used in this book are as follows:

DIAMETER	⌀	⌀	⌀
CONICAL TAPER	▷	▷	▷
COUNTERBORE/SPOTFACE	⊔	⊔ (PROPOSED)	⊔
COUNTERSINK	∨	∨ (PROPOSED)	∨
DEPTH/DEEP	↧	↧ (PROPOSED)	↧
RADIUS	R	R	R

Counterbores, spotfaces, and countersinks can be shown on drawings by dimension symbols or abbreviations with the symbols being preferred. Some samples of these symbols in use are shown in Figures A and B.

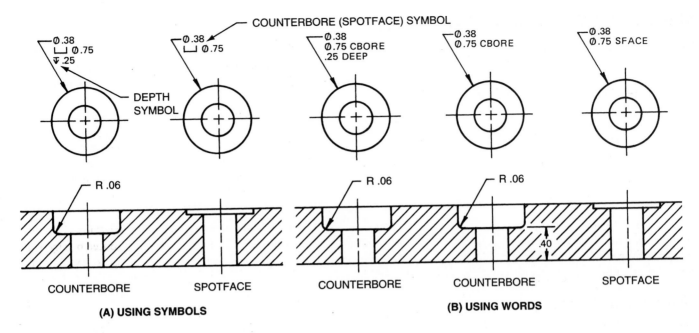

(A) USING SYMBOLS

(B) USING WORDS

FIG. A COUNTERBORED AND SPOTFACED HOLES

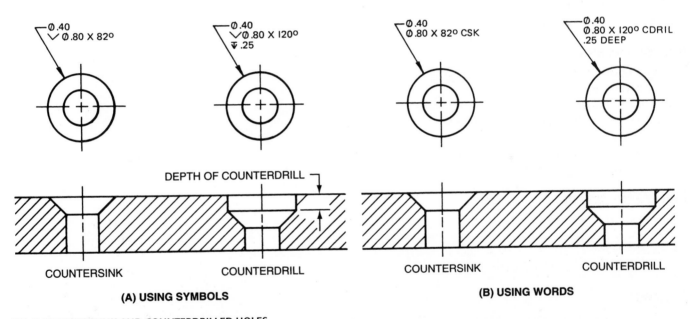

(A) USING SYMBOLS

(B) USING WORDS

FIG. B COUNTERSUNK AND COUNTERDRILLED HOLES

Cecil Jensen and Jay D. Heisel, *Engineering Drawing and Design*, 5th edition. Glencoe/McGraw-Hill, 1996, pp. 2–45, 182, & 183.

SECTION 1

TESTS 1-91

TEST 1 Safety *(Unit 4)*

Most accidents in a machine shop could be avoided as they are usually caused by carelessness, thoughtlessness, or even recklessness on someone's part.

Safety programs developed by industry, the safety council, or accident prevention associations are meaningless unless they are practiced by EVERYONE in the shop. It is the duty and responsibility of every worker to develop safe work habits.

Place the correct word(s) in the blank space(s) provided at the right-hand side of the page that will make the sentence complete and true.

1. Accidents can be avoided if workers develop __?__ work habits.

    1. _safe_

2. A safe worker should always think of his or her own safety and that of __?__ workers.

    2. _other_

3. Safety glasses with __?__ shields offer good eye protection.

    3. _side_

4. When lifting heavy objects, keep your __?__ bent and your back __?__ and lift with __?__ muscles.

    4. _knees_
    straight
    leg

5. The machine should be __?__ before cleaning or oiling it.

    5. _stopped_

6. Before repairing a machine, disconnect the __?__ and lock it off at the __?__ box.

    6. _power_
    switch

7. Be sure that the safety __?__ are in __?__ before operating any machine.

    7. _device_
    working order

Decide whether each statement is true or false and circle the letter in the right-hand column that indicates your choice.

8. Never wear loose clothing when operating a machine.

    8. **(T)** F

9. Canvas shoes or sandals should not be worn in a shop.

    9. **(T)** F

10. Compressed air can be used to clean machines.

    10. T **(F)**

11. Always use a brush to remove cuttings from a machine.

    11. **(T)** F

12. Revolving work may be measured safely.

    12. T **(F)**

13. It is dangerous for a machinist to have long, unprotected hair when working on machine tools.

    13. **(T)** F

14. A good machinist does not have to be concerned with good safety practices.

    14. T **(F)**

15. Never wear watches or rings in a machine shop.

    15. **(T)** F

16. Minor cuts or injuries do not have to be treated immediately.

    16. T **(F)**

SCORE _____

20

TEST 2 Engineering Drawings *(Unit 5)*

Engineering drawings are very important in any trade as they convey valuable information about a product and its dimensions to a worker. The machinist must understand how to interpret these drawings as they contain all the necessary information for producing the part.

Place the correct word(s) in the blank space(s) provided at the right-hand side of the page that will make the sentence complete and true.

1. Orthographic projection shows a view of the part from __?__ sides.
2. Cylindrical parts are generally shown with __?__ views.
3. The __?__ and __?__ permissible dimensions of a part are called the *limits*.
4. The permissible variation of a part size is called the __?__.
5. The intentional difference in mating parts is called the __?__.
6. A view that uses an imaginary line to expose the interior or the contour form of a part is called a __?__ view.

1. _3_
2. _2_
3. _largest_
   smallest
4. _tolerance_
5. _Allowance_
6. _sectional_

Identify the following lines used in enginering drawings and place your answers in the right-hand column.

7. _____
8. - - - - - - - - - - - - - - - -
9. ___ __ __ ___ __ __ ___
10. |————————————————|

7. _Object_
8. _Hidden_
9. _cutting-plane_
10. _dimension_

Name one material indicated by each of the following symbols and place your answers in the right-hand column.

11.
12.
13.
14.

11. _copper_
12. _aluminum_
13. _steel_
14. _cast iron_

Identify the common machine shop abbreviations or symbols found on an engineering drawing and place your answers in the right-hand column.

15. *R*
16. Csk
17. Ø
18. hdn
19. mm

15. _Radius_
16. _Countersink_
17. _diameter_
18. _Harden_
19. _millimeter_

SCORE _____

20

TEST 3 Machining Procedures for (Unit 6)
Various Workpieces

The importance of following the correct sequence to machine a part cannot be overemphasized. If the correct sequence for machining is not followed, valuable time may be wasted or the part may be ruined or damaged.

Place the correct word(s) in the blank space(s) provided at the right-hand side of the page that will make the sentence complete and true.

1. When machining round work, rough-turn all diameters to within __?__ of the required size.

    1. _.030"_

2. Machine the __?__ diameter first and proceed to the __?__ .

    2. _largest_
       smallest

3. All steps and shoulders should be rough-turned to within __?__ of the length required.

    3. _.030"_

4. When machining steps or shoulders, take all measurements from the __?__ of the workpiece.

    4. _End_

5. Metal expands due to the __?__ caused by the machining process.

    5. _friction_

6. When machining work in a chuck, hold the workpiece __?__ for rigidity.

    6. _short_

7. Work should never extend more than __?__ times its diameter beyond the chuck jaws.

    7. _three_

8. Long work held in a chuck should be supported by a __?__ rest or __?__ .

    8. _steady_
       center

9. When cutting an internal thread, cut a groove at the end of the thread slightly deeper than the __?__ diameter.

    9. _major_

10. When a finished diameter is held in a chuck, the surface should be protected with a piece of __?__ metal.

    10. _soft_

11. The diameter of a workpiece will be __?__ if measured while it is hot.

    11. _incorrect_

12. The center of all holes to be drilled should be __?__ punched.

    12. _prick_

13. Rough stock should be cut a little __?__ than the finished length.

    13. _longer_

Select the most appropriate answer for each question and circle the letter in the right-hand column that indicates your choice.

14. All layout lines that indicate surfaces to be cut should be
    (A) scribed very lightly   (C) prick-punched
    (B) very wide   (D) center-punched

    14.   A   B   (C)   D

15. When cutting out a section of a workpiece on a bandsaw, prior to further machining you should
    (A) cut to the layout line   (C) use a wide blade
    (B) cut .12 in. outside the layout line   (D) cut .03 in. outside the layout line

    15. (A)   B   C   D

16. The first surface to be machined on a flat workpiece should be
    (A) the smallest   (C) the edge
    (B) the largest   (D) the back

    16.   A   B   (C)   D

17. How much material should be left on each surface to be ground?
    (A) .03 in.   (C) .010 in.
    (B) 2 mm   (D) .002 in.

    17.   A   B   (C)   D

18. The surface of a workpiece being milled should
    (A) be level   (C) be filed first
    (B) project at least .12 in. above the top of   (D) project at least .75 in. above the top
        the vise   of the vise

    18.   A   B   (C)   D

SCORE _____

TEST 4 Basic Measurement *(Unit 7)*

One of the most important phases of machine shop work is that of measurement. The student must be familiar with the most basic measuring instrument, that is, the rule, before proceeding to use the more precise types of measuring equipment.

Decide whether each statement is true or false and circle the letter in the right-hand column that indicates your choice.

1. If the end of a rule is worn, the measurement should be taken from the 1-in. or 1-cm line. 1. (T) F

2. Rules may be used to check the flatness of a workpiece. 2. (T) F

3. Two types of calipers are the spring-joint and the flexible-joint caliper. 3. T (F)

4. When setting an outside caliper to a rule, place one leg of the caliper over the end of the rule. 4. (T) F

5. Revolve the work in a lathe slowly when measuring with a caliper. 5. (T) F

6. Spring-joint calipers may be used for making fine measurements. 6. T (F)

7. The work is the correct diameter when the outside caliper slides over the work by its own weight. 7. (T) F

8. When checking the setting of an inside caliper, place the end of the rule and one caliper leg against a flat surface. 8. (T) F

9. To properly set inside or outside calipers, hold the ends of the legs parallel to the edge of the rule. 9. T (F)

10. If an accurate measurement is required, the inside-caliper setting should be checked with an outside caliper. 10. T (F)

11. **Record the rule reading for each illustration in the proper space in the right-hand column.**

A

B

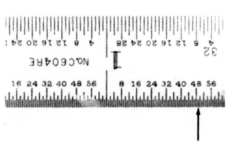

C

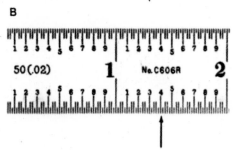

D

11. A _15/64 in_

    B _1 11/64_

    C _1 49/64_

    D _1.4_

    E _32 mm_

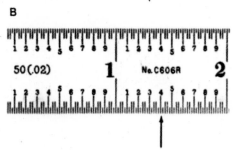

E

continued on next page

12. The common fractions found *in order* from smallest to largest on a #4 graduations steel rule are ___?___ , ___?___ , ___?___ , and ___?___ .

12.   ¹/₈
       ¹/₁₆
       ¹/₃₂
       ¹/₆₄

13. The common graduations found on a decimal rule are .100, .020, .050, and ___?___ .

13.   .010

SCORE _____
                    20

TEST 5 Squares and Surface Plates *(Unit 8)*

Squares and surface plates are used for layout, inspection, and setup purposes. These tools are really the basic tools for these operations because with them all horizontal and vertical reference surfaces or lines are produced.

Place the correct word(s) in the blank space(s) provided at the right-hand side of the page that will make the sentence complete and true.

1. Precision squares are hardened and accurately __?__ and must be handled carefully to __?__ their accuracy.

   1. *ground*
      preserve

2. Squares may be of the __?__ or __?__ type.

   2. *solid*
      adjustable

3. Better-quality inspection squares have a __?__-edge blade that makes a __?__ contact with the work.

   3. *beveled*
      line

4. __?__ will show through when the blade does not contact the work for its full length.

   4. *light*

5. Work may be checked for squareness on a surface plate by using a square and two strips of __?__ between the work and the __?__ .

   5. *paper*
      square

6. Cast-iron surface plates are well ribbed and supported to resist __?__ under heavy loads.

   6. *deflection*

7. Surface plates are made of __?__-grained cast iron, which has good __?__-resistance qualities.

   7. *close*
      wear

8. Granite surface plates are manufactured from gray, __?__ , or __?__ granite.

   8. *gray*
      pink

9. Extremely flat finishes on granite plates are produced by __?__ .

   9. *lapping*

Select the most appropriate answer for each question and circle the letter in the right-hand column that indicates your choice.

10. Which of the following statements does not apply to the care of a square?
    (A) Never drop a square.
    (B) Keep it clean at all times.
    (C) Apply a light film of oil before use.
    (D) Store a square in a special box.

    10. A   B   (C)   D

11. To prevent rocking on an uneven surface, a surface plate has
    (A) three-point suspension
    (B) a wedge for leveling
    (C) four-point suspension
    (D) adjustable legs

    11. (A)   B   C   D

12. The surface of a cast-iron surface plate is finished by
    (A) grinding
    (B) filing
    (C) scraping
    (D) lapping

    12. A   B   (C)   D

13. Which of the following statements does not apply to granite surface plates?
    (A) They are available in several grades.
    (B) They are appreciably affected by temperature changes.
    (C) They are rustproof.
    (D) They are nonmagnetic.

    13. A   (B)   C   D

14. Granite surface plates are considered better than cast-iron plates because they
    (A) are lighter
    (B) do not chip
    (C) do not burr
    (D) all of the previous

    14. A   B   (C)   D

SCORE _____

20

TEST 6 Micrometers *(Unit 9)*

The most common precision outside-measuring instrument is the micrometer. It is probably the most used measuring instrument in the machine shop. Micrometers are generally capable of measuring to an accuracy of .001 or .0001 in.

Select the most appropriate answer for each question and circle the letter in the right-hand column that indicates your choice.

1. On the micrometer spindle there are
   (A) 1000 threads per inch
   (B) 25 threads per inch
   (C) 40 threads per inch
   (D) 100 threads per inch

   1. A   B   Ⓒ   D

2. One complete revolution of the micrometer thimble increases or decreases the distance between measuring faces by
   (A) .025 in.
   (B) .001 in.
   (C) .100 in.
   (D) .0001 in.

   2. Ⓐ   B   C   D

3. If the thimble is revolved counterclockwise from the fully closed position so that three lines are showing on the sleeve, the distance between measuring faces would be
   (A) .300 in.
   (B) .075 in.
   (C) .003 in.
   (D) .750 in.

   3. A   Ⓑ   C   D

4. A #6 showing on the sleeve would indicate a distance between measuring faces of
   (A) .600 in.
   (B) .150 in.
   (C) .006 in.
   (D) .015 in.

   4. Ⓐ   B   Ⓒ   D

5. Forty divisions on the sleeve of a micrometer are equal to
   (A) .400 in.
   (B) .040 in.
   (C) 1.000 in.
   (D) .100 in.

   5. A   Ⓑ   C   D

6. Each division around the thimble of a micrometer is equal to
   (A) .100 in.
   (B) .010 in.
   (C) .001 in.
   (D) .0001 in.

   6. A   B   Ⓒ   D

7. The value of each vernier division equals
   (A) .001 in.
   (B) .0005 in.
   (C) .0001 in.
   (D) .005 in.

   7. A   B   Ⓒ   D

8. A vernier micrometer has
   (A) two scales on the thimble
   (B) two scales on the sleeve
   (C) an indicating dial
   (D) a sliding vernier bar

   8. A   Ⓑ   C   D

9. The vernier scale on a vernier micrometer consists of
   (A) 10 divisions
   (B) 100 divisions
   (C) 1000 divisions
   (D) 10,000 divisions

   9. Ⓐ   B   C   D

Place the correct word(s) in the blank space(s) provided at the right-hand side of the page that will make the sentence complete and true.

10. When checking the micrometer for accuracy, close the measuring __?__ by turning the __?__ .

    10. faces
        Ratchet

11. The indicating micrometer may be used as a __?__ by setting it to the desired size with gage blocks.

    11. comparator

12. Two basic differences between a standard micrometer and a metric micrometer are the pitch of the spindle screw and the __?__ on the thimble.

    12. graduations

13. On the metric micrometer, the circumference of the thimble is divided into __?__ equal divisions.

    13. 50

14. The pitch of a metric micrometer screw is __?__ mm.

    14. 0.5

15. Every __?__ graduation on the metric micrometer sleeve is numbered.

16. The __?__ - __?__ micrometer has graduations on the thimble and barrel as in a standard micrometer along with a __?__ readout built into the frame.

17. The indicating micrometer uses an __?__ dial and a movable __?__ to permit accurate measurements to ten-thousandths of an inch (0.002 mm).

18. The Digi-Matic micrometer with __?__ - __?__ control provides a stand-alone inspection system that can be interfaced with a computer.

15. *fifth*

16. *direct-Reading*
    digital

17. *dictating*
    anvil

18. *statistical process*

Record the micrometer reading for each illustration in the proper space in the right-hand column. State whether the reading is in. or mm.

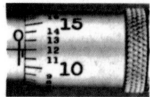

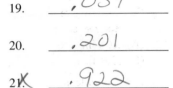

19.

20.

21.

19. *.037*

20. *.201*

21.✗ *.922*
    .898

22.

23.

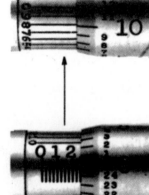
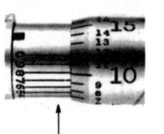
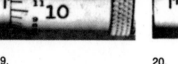

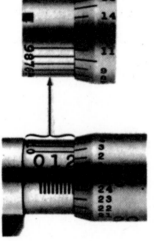

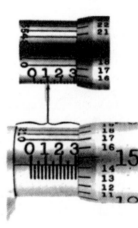

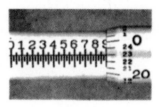

22.

23.

24.

24.

25.

26.

27.

SCORE

30

25.

26.

27.

TEST 7 Vernier Calipers *(Unit 10)*

The vernier caliper, capable of measuring to .001 in. or 0.02 mm, has a much wider range than the micrometer. It may be used for making inside and outside measurements equal to and beyond the range of micrometers.

Place the name of each illustrated vernier caliper part in the proper space in the right-hand column.

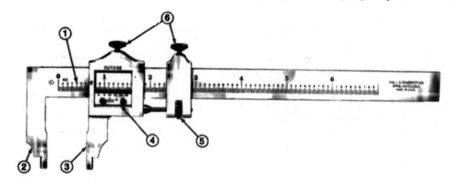

1. _Bar_
2. _Fixed jaw_
3. _movable jaw_
4. _vernier scale_
5. _fine adjust. nut_
6. _clamp screws_

Select the most appropriate answer for each question and circle the letter in the right-hand column that indicates your choice.

7. On a 25-division vernier, 25 divisions on the vernier scale are equal in length to
   (A) 1.000 in.   (C) .600 in.
   (B) .100 in.   (D) .060 in.

   7. A   B   (C)   D

8. On a 25-division vernier, 25 divisions on the vernier scale are equal to
   (A) 10 divisions on the bar   (C) 12 divisions on the bar
   (B) 100 divisions on the bar   (D) 24 divisions on the bar

   8. A   B   C   (D)

9. On a 50-division vernier, 50 divisions on the vernier scale are equal to
   (A) 49 divisions on the bar   (C) 10 divisions on the bar
   (B) 50 divisions on the bar   (D) 100 divisions on the bar

   9. (A)   B   C   D

10. The reason for a 50-division vernier caliper is to provide
    (A) easier reading   (C) a metric conversion
    (B) greater accuracy   (D) a metric calculation

    10. (A)   B   C   D

11. The small numbers on the bar of a 25-division vernier caliper indicate
    (A) .100 in.   (C) .025 in.
    (B) 1.000 in.   (D) .050 in.

    X 11. (A)   B   (C)   D

12. In actual measurement, each division on the 25-division vernier scale is equal to
    (A) .010 in.   (C) .024 in.
    (B) .100 in.   (D) .050 in.

    12. A   B   (C)   D

13. In actual measurement, each division on the 50-division vernier scale is equal to
    (A) .049 in.   (C) .100 in.
    (B) .050 in.   (D) .010 in.

    13. (A)   B   C   D

Record the vernier reading for each illustration in the proper space in the right-hand column.

VERNIER CALIPER READINGS (25 GRADUATIONS)

X14. _1.001_

15. _3.066_

14.   15.

16.

17.

VERNIER CALIPER (50 GRADUATIONS)

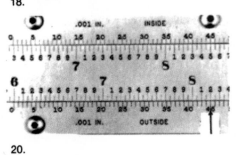

18.

19.

18. 4.602

19. 1,318

5.995 X 20. 6.045

21. .012

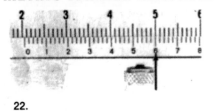

20.

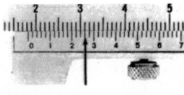

21.

METRIC VERNIER READINGS

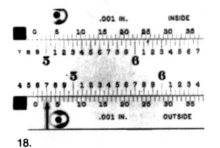

22.

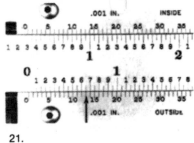

23.

24.

25.

22. _____

23. _____

24. _____

25. _____

SCORE _____

25

TEST 8 Inside-Measuring Instruments *(Unit 11)*

The ability to accurately measure an inside surface is far more demanding than that required to measure an outside surface. Most inside-measuring instruments depend on the operator's sense of "feel" for making accurate and consistent measurements.

Decide whether each statement is true or false and circle the letter in the right-hand column that indicates your choice.

1. The inside micrometer caliper can measure hole diameters from .100 to 2.000 in.   1.   T   (F)
2. Some inside micrometer calipers have reverse barrel readings.   2.   T   F
3. Inside micrometers are transfer-type instruments.   3.   T   F
4. A locking nut is not required on an inside micrometer since the spindle nut is a tight fit on the spindle thread.   4.   T   F
5. Inside micrometers are available to measure from 1.500 to 100 in.   5.   T   F
6. Inside micrometers sometimes require a spacing collar to obtain the proper measurement.   6.   T   F
7. An Intrimik has four measuring faces for greater accuracy.   7.   T   F
8. The Intrimik is a transfer-type measuring instrument.   8.   T   F
9. Small hole gages are manufactured in two styles—the rounded type and the flat-bottom type.   9.   T   F
10. The rounded-type small hole gage is used for measuring shallow slots and holes.   10.   T   F
11. A telescoping gage is locked to a size by turning the handle.   11.   T   F
12. A dial bore gage may be used to check a hole for both taper and depth.   12.   T   F

Place the name of each illustrated instrument in the proper space in the right-hand column.

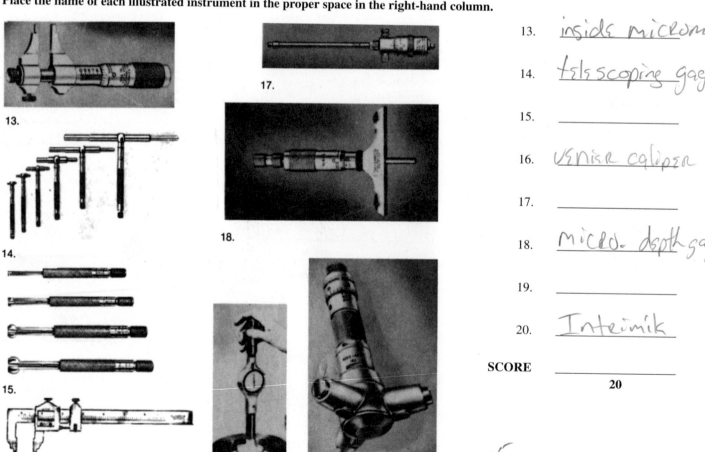

13.

14.

15.

16.

17.

18.

19.   20.

13. _inside micrometer_

14. _telescoping gages_

15. _____

16. _venier caliper_

17. _____

18. _micro. depth gage_

19. _____

20. _Intrimik_

SCORE _____

20

.6875
.005
.5870
.1370
.456

TEST 9 Depth and Height *(Unit 11)* Measurement

Although rules and various attachments may be used for depth measurement, the depth micrometer and the depth vernier are generally used where accuracy is required.

The vernier height gage is probably the most commonly used height-measuring instrument and is widely used for layout and inspection purposes. With a depth-gage attachment, the vernier height gage may also be used for checking the depth of holes, slots, etc.

Place the correct word(s) in the blank space(s) provided at the right-hand side of the page that will make the sentence complete and true.

1. Depth micrometers may have a range of .000 to __?__, which is achieved by means of __?__ rods.

    1. _____

        _____

2. Before measuring the depth of a hole, remove all __?__ from the edge of the hole.

    2. _____

3. When taking a depth measurement, rotate the thimble of the depth micrometer in a __?__ direction using the tip of one __?__ .

    3. _____

        _____

4. The accuracy of the depth gage is controlled by a __?__ on the end of each rod.

    4. _____

5. A vernier height gage can be accurately set to within __?__ .

    5. _____

6. A vernier height gage can be used for accurate layout when a __?__ is attached to the movable jaw.

    6. _____

7. An offset scriber allows a height gage to be set from the face of the __?__ __?__ .

    7. _____

        _____

8. If a dial indicator is used with a vernier height gage, an accuracy of __?__ can be obtained.

    8. _____

9. If greater accuracy than that possible with a vernier height gage and dial indicator is required, __?__ blocks may be used.

    9. _____

10. When using a precision height gage, use a surface plate as a __?__ surface.

    10. _____

11. The distance between the steps on a precision height gage is exactly __?__ .

    11. _____

12. The column on a precision height gage may be raised or lowered __?__ .

    12. _____

13. The range of a precision height gage may be increased by the use of a __?__ block.

    13. _____

Record the micrometer-depth-gage reading for each illustration in the proper space in the right-hand column.

14. _____

15. _____

16. _____

17. _____

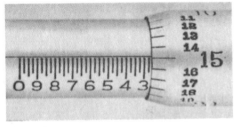

14.

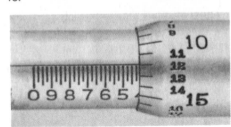

15.

16.

17.

SCORE _____

20

TEST 10   Gage Blocks *(Unit 12)*

Gage blocks are the accepted standard of accuracy in the manufacturing sector. Their accuracy provides a means of checking and maintaining high standards of measurement and has contributed to high production and interchangeable manufacture.

Select the most appropriate answer for each question and circle the letter in the right-hand column that indicates your choice.

1. A gage block is only accurate when measured at
   (A) 32°F   (C) 68°F
   (B) 0°F   (D) 100°F

   1. A   B   C   D

2. Gage blocks are used to check the dimensional accuracy of fixed gages to determine the extent of
   (A) wear   (C) shrinkage
   (B) growth   (D) all of the previous

   2. A   B   C   D

3. The purpose of wear blocks is to
   (A) provide an additional .050- or   (C) prolong the life and accuracy of the
       .100-in. block   remainder of the set
   (B) act as substitutes for the most   (D) lap-test surfaces accurately for
       commonly used gage blocks   subsequent gaging

   3. A   B   C   D

4. It is a good habit to always expose the same face of the wear block to the work surface because
   (A) any inaccuracy will be unilateral (C) the accuracy life will be doubled
   (B) the wringing quality of the opposite (D) known inaccuracies can be easily
       face will be preserved   recognized

   4. A   B   C   D

5. Gage blocks are manufactured in sets. The working set is
   (A) AA   (C) C
   (B) B   (D) A

   5. A   B   C   D

6. The best accuracy can be obtained when, prior to use, both gage blocks and work are placed in
   (A) CO_2 (C) kerosene
   (B) machine oil   (D) any noncorrosive fluid

   6. A   B   C   D

7. To remove dust particles and to apply a thin film of oil to the blocks before wringing them together, it is recommended that they be
   (A) placed in an air jet   (C) wiped with the palm of the hand
   (B) rubbed on leather   (D) immersed in oil

   7. A   B   C   D

8. *Without the use of wear blocks,* the least number of gage blocks needed from an 83-piece set to make a 2.7005-in. buildup would be
   (A) two   (C) four
   (B) three   (D) five

   8. A   B   C   D

9. *With the use of .100-in. for wear blocks,* the least number of gage blocks needed from an 83-piece set to make a 10.9999-in. buildup would be
   (A) seven   (C) nine
   (B) eight   (D) ten

   9. A   B   C   D

10. In the .0001-in. series of an 83-piece gage block set, the sizes would range from
    (A) .1000 to .1010 in.   (C) .1001 to .1009 in.
    (B) .0001 to .0009 in.   (D) .1010 to .1019 in.

    10. A   B   C   D

Decide whether each statement is true or false and circle the letter in the right-hand column that indicates your choice.

11. Gage blocks are manufactured only in 115- and 83-piece sets.

    11. T   F

12. Gage block measuring surfaces are lapped to within a few millionths of an inch.

    12. T   F

13. When long wear is desirable, carbide blocks should be used.

    13. T   F

14. A "standard-inch" size is accurate only when measured at a standard temperature of 100°F.

      14.   T   F

15. Wear blocks are provided in .100-in. size only.

      15.   T   F

16. A good habit to formulate is to use both sides of the wear blocks alternately to minimize wear.

      16.   T   F

17. A Class B set is commonly called a *working set.*

      17.   T   F

18. Gage blocks, made of hardened alloy steel, are not affected by temperature change.

      18.   T   F

Place the correct word(s) in the blank space(s) provided at the right-hand side of the page that will make the sentence complete and true.

19. Gage blocks have made __?__ manufacture possible.

      19. _____

20. During manufacture, gage blocks are __?__ to remove strain.

      20. _____

21. Gage blocks are accurate to size only at __?__ °F.

      21. _____

22. Gage blocks are used to check the dimensional __?__ of fixed gages.

      22. _____

23. The Class __?__ set is used for inspection purposes.

      23. _____

24. The laboratory or master set of gage blocks is accurate to within __?__ of an inch.

      24. _____

25. If a temperature-controlled room is not available, both the work and gage blocks should be placed in __?__ until both are at the same temperature.

      25. _____

              SCORE _____

                                          25

TEST 11   Angular Measurement *(Unit 13)*

The bevel protractor is commonly used in machine shop work for laying out, checking, and setting up angular surfaces. When more accuracy is required, a vernier bevel protractor or a sine bar and gage blocks may be used.

Select the most appropriate answer for each question and circle the letter in the right-hand column that indicates your choice.

1. A quadrant on a universal bevel protractor consists of
   (A) 45°   (C) 90°
   (B) 50°   (D) 100°

   1. A   B   C   D

2. The universal bevel protractor has an accuracy of
   (A) 5′   (C) 0.5°
   (B) 5″   (D) 5°

   2. A   B   C   D

3. A universal bevel protractor can measure any angle up to
   (A) 90°   (C) 270°
   (B) 180°   (D) 360°

   3. A   B   C   D

4. On a universal bevel protractor, the difference between one vernier division and two main-scale divisions represents
   (A) 0.5°   (C) 1°
   (B) 5′   (D) 23′

   4. A   B   C   D

5. Twelve vernier-scale spaces occupy the same space on the main scale as
   (A) 11°   (C) 23°
   (B) 22°   (D) 24°

   5. A   B   C   D

6. A sine bar must be used to check an angle of
   (A) 5′   (C) 5 to 10′
   (B) more than 10′   (D) less than 5′

   6. A   B   C   D

7. The 5-in. distance on a sine bar is measured
   (A) by the length of the bar   (C) between the centers of the cylinders
   (B) between the inside diameters of the   (D) between the outside diameters of the
       cylinders   cylinders

   7. A   B   C   D

8. In theory, the sine bar becomes the
   (A) hypotenuse of a right-angle triangle   (C) side adjacent of a right-angle triangle
   (B) side opposite of a right-angle triangle (D) right angle of a right-angle triangle

   8. A   B   C   D

9. When an angle greater than 60° is to be checked, it is better to set up the work using
   (A) an angular converter   (C) the supplementary angle
   (B) the complement of the angle   (D) the converted angle

   9. A   B   C   D

10. When the sine bar buildup required is smaller than the available gage block size, the alternative method is to use
    (A) a bevel protractor   (C) the net difference between two gage
    (B) the complementary angle   block buildups
                                     (D) the supplementary angle

    10. A   B   C   D

    SCORE _____
                    10

TEST 12　　Fixed Gages　*(Unit 14)*

Fixed gages provide a means of quickly and accurately checking the dimensions of a part. These gages are generally of the "go" or "no-go" variety and are used to check the limits of the dimensions but not the actual size of the part.

Place the correct word(s) in the blank space(s) provided at the right-hand side of the page that will make the sentence complete and true.

1. On a cylindrical plug gage, the smaller-diameter plug is called the __?__ gage and it measures the __?__ limit of the hole.

2. The larger diameter of the plug checks the __?__ limit of the hole or the __?__ size permissible.

3. When using a plug gage, always check the gage and the __?__ for nicks and __?__ .

4. Before use, wipe the ends of the gage with an __?__ cloth.

5. Always start a plug gage __?__ in the hole.

6. If the hole being checked is within the __?__ , the __?__ gage will enter easily.

7. A ring gage is ground and __?__ internally to the desired size.

8. A "no-go" ring gage is identified by an __?__ ring on the knurled surface.

9. If the hole taper is not correct, the taper plug gage will __?__ in the hole.

10. When checking a tapered hole, apply a __?__ coating of Prussian blue to the surface of the plug gage.

11. When checking a tapered hole, rotate the plug gage __?__ for only one- __?__ turn.

12. If the gage is a proper fit in the hole, the bluing will rub off for the __?__ of the gage.

13. When checking an unfinished external taper, apply three __?__ spaced __?__ lines around the circumference of the workpiece.

14. When checking an internal thread with a gage, always apply a little __?__ and never __?__ the gage into the thread.

15. A thread ring gage has a __?__ screw to permit small adjustment.

16. A snap gage is used to check the diameter of a part to the preset __?__ of the snap gage.

17. Snap gages may have anvils or __?__ for checking the work size.

18. Dial- __?__ snap gages will give a direct reading of the part size.

1. _____
   _____
2. _____
   _____
3. _____
   _____
4. _____
5. _____
6. _____
   _____
7. _____
8. _____
9. _____
10. _____
11. _____
   _____
12. _____
13. _____
   _____
14. _____
   _____
15. _____
16. _____
17. _____
18. _____

SCORE _____

25

TEST 13   Comparison Measurement (1) *(Unit 15)*

Comparison Measurement is the process of comparing the part size to a known standard. This form of measurement uses some form of amplification to indicate the part size, which makes it easy and quick to see whether the dimensions fall within the required limits for the part.

Place the correct word(s) in the black space(s) provided at the right-hand side of the page that will make the sentence complete and true.

1. Dial indicators may be used to compare the dimensions of a workpiece to a known __?__ and to check the __?__ of a workpiece prior to machining.

2. Many types of dial indicators operate on the gear and __?__ principle.

3. Dial indicators are generally of two types, which are the __?__-reading indicator and the __?__ test indicator.

4. Regular-range dial indicators have only about __?__ revolutions of the needle.

5. The long-range dial indicator usually has about one __?__ of travel.

6. __?__ dial test indicators have the plunger at right angles to the dial.

7. A __?__ dial test indicator may be used to check internal and external surfaces.

8. Mechanical comparators may operate on the __?__ and rack principle or use a system of __?__.

9. Optical comparators are suited for checking the dimensions of __?__ and __?__-shaped parts.

10. Electronic gages are suited to checking soft, highly polished surfaces because of the __?__ gaging pressure required.

11. The mechanical-optical comparator combines a __?__ mechanism with a light beam to check the part size by means of a __?__ cast on a magnified scale.

12. The reed mechanism consists of a fixed and a __?__ block to which two pieces of __?__ steel, called *reeds,* are attached.

13. The reed-comparator spindle is set to a __?__ gage prior to measuring a workpiece.

14. A small movement of the spindle on a reed comparator produces a larger movement of the __?__, which is attached to the fixed and movable blocks.

1. _____
   _____
2. _____
3. _____
   _____
4. _____
5. _____
6. _____
7. _____
8. _____
   _____
9. _____
   _____
10. _____
11. _____
    _____
12. _____
    _____
13. _____
14. _____

SCORE _____

20

TEST 14   Comparison Measurement (2) *(Unit 15)*

Air gages are used extensively in industry because they have certain advantages over other types of gages. Some of these are:
- They can check holes for out-of-roundness, taper, and concentricity.
- The gage will not mar the finish.
- Less operator skill is required in their use.
- They can check more than one diameter at a time.

Place the correct word(s) in the blank space(s) provided at the right-hand side of the page that will make the sentence complete and true.

1. The __?__ or column-type gage measures the part size by air __?__ .

2. The pressure-type gage measures the part size by air __?__ .

3. In the column type, the air flows through a transparent __?__ tube that is attached to the gaging __?__ by means of a plastic tube.

4. In the column type, the air flows through passages in the __?__ head and the size of the hole is indicated by the height of a __?__ that is controlled by the amount of airflow between the head and the __?__ .

5. If the hole is oversize, the float will __?__ in the tube.

6. If the hole is too small, the float will __?__ in the tube.

7. In a pressure-type air gage, the air is divided into two channels. These are the __?__ channel and the __?__ channel.

8. These two channels are connected with an extremely accurate differential __?__ meter.

9. To set the pressure-type gage, place a __?__ over the __?__ spindle and adjust the needle to zero with the __?__ __?__ valve.

10. If the hole is too __?__ , more air will escape through the gaging plug and the pressure in the gaging channel will be __?__ .

11. Air-type gaging heads last longer than others because __?__ is reduced between the gaging head and the workpiece.

1. _____
   _____

2. _____

3. _____
   _____

4. _____
   _____
   _____

5. _____

6. _____

7. _____
   _____

8. _____

9. _____
   _____
   _____
   _____

10. _____
    _____

11. _____

SCORE _____

20

TEST 15   The Coordinate Measuring (*Unit 16*) System

Coordinate measurement is a method by which all measurements in one direction are made from one surface. This method for layout, machining, and checking is used with the aid of some accurate measuring devices.

Select the most appropriate answer for each question and circle the letter in the right-hand column that indicates your choice.

1. Coordinate measuring systems can measure
   (A) on the X and Y axes
   (B) to an accuracy of .000010 in.
   (C) both A and B
   (D) neither A nor B

   1.   A   B   C   D

2. Which component is *not* part of a coordinate measuring unit?
   (A) spar
   (B) reading head
   (C) analog display
   (D) digital display

   2.   A   B   C   D

3. The transparent index grating of the measuring unit
   (A) has fewer graduations per inch than the spar
   (B) has its lines at an angle to those on the spar
   (C) has lines parallel to the spar lines
   (D) has finer graduations than those of the spar

   3.   A   B   C   D

4. The lines on the spar and index gratings are produced by
   (A) stamping
   (B) machining
   (C) scraping
   (D) etching

   4.   A   B   C   D

5. A collimating lens converts light from the lamp into a
   (A) parallel beam
   (B) diverging beam
   (C) converging beam
   (D) wide-angle beam

   5.   A   B   C   D

6. As the reading is moved longitudinally along the face of the spar, the fringe pattern
   (A) moves vertically
   (B) is converted into an electrical signal
   (C) both A and B
   (D) neither A nor B

   6.   A   B   C   D

7. Which of the following statements is *not true* in regard to coordinate measuring systems?
   (A) They reduce setup time.
   (B) They reduce machining time.
   (C) They reduce floor-to-floor time for the part.
   (D) They are easy to read.

   7.   A   B   C   D

8. When using a coordinate measuring system, one
   (A) requires more skill
   (B) requires a good knowledge of mathematics
   (C) must make several calculations before machining
   (D) reduces the chance of error

   8.   A   B   C   D

9. The coordinate measuring system
   (A) requires the use of gage blocks
   (B) removes the need to check dimensions during machining
   (C) requires the use of measuring rods
   (D) none of the previous

   9.   A   B   C   D

10. When the light beam strikes the fringe pattern, it is reflected
    (A) to a photoelectric cell
    (B) back to the spar
    (C) to the digital-readout box
    (D) none of the previous

    10.   A   B   C   D

SCORE _____

10

TEST 16   Measuring with Light Waves *(Unit 17)*

Although there are many precision measuring tools and devices, one of the most accurate forms of measurement involves the use of light waves. This principle is used in extremely accurate measuring procedures involving optical flats, laser devices, and the interferometer.

Place the correct word(s) in the blank space(s) provided at the right-hand side of the page that will make the sentence complete and true.

1. Optical flats are used with a __?__ light.   1. _____

2. Optical flats are disks of clear, fused __?__ .   2. _____

3. Helium light has a wavelength of 23.1323 __?__ of an inch.   3. _____

4. When two light waves cross each other, they are said to __?__ .   4. _____

5. Each band that is visible when an optical flat is used represents __?__ microinches. 5. _____

6. When a part is checked against a master, the part is __?__ when pressure applied on the master makes the band spacing wider.   6. _____

7. An interferometer checks machine part alignment by means of light-wave __?__ .   7. _____

8. The main parts of an interferometer are the beam splitter, the motion-sensitive mirror, and the __?__ .   8. _____

9. The beam produced by an interferometer is very thin and __?__ .   9. _____

10. When an interferometer is being used and the two beams are out of phase, the light fluctuations are computed and the precise movement is displayed on a __?__ box.   10. _____

SCORE _____
                                                                10

TEST 17   Surface-Finish Measurement *(Unit 18)*

Surface finish has become an important part of manufacturing and assembly processes. Surface finish is critical where it is necessary to minimize wear and friction between assembled parts. High surface finishes are found in automobile cylinder walls, bearings, hydraulic cylinders, etc. Precision surface finishes may be produced by grinding, honing, and lapping.

Place the correct word(s) in the blank space(s) provided at the right-hand side of the page that will make the sentence complete and true.

1. The desired surface finish is indicated on the print to prevent __?__ of the part.

2. The inch unit of surface-finish measurement is the __?__ .

3. The two main parts of a surface-finish indicator are the __?__ head and the __?__ .

4. The surface-indicator meter reading indicates the average __?__ of surface roughness.

5. Any departure from the nominal surface, such as waviness, roughness, and flaw, is defined as surface __?__ .

6. A __?__ pattern deviates from the mean surface in the form of waves. It is usually caused by __?__ in the machine or the work.

7. The finely spaced irregularities caused by the cutting-tool action and superimposed on the waviness pattern are called __?__ .

8. Surface irregularities such as scratches and holes are called __?__ .

9. The direction of the predominant surface pattern is called the __?__ .

10. The arithmetical average deviation measured normal to the centerline in microinches is called the roughness __?__ .

11. The symbol __?__ is used to indicate a lay pattern that is parallel to the indicated surface.

12. A multidimensional pattern is indicated by the symbol __?__ .

13. The symbol C indicates a lay pattern that is approximately __?__ to the center of the surface indicated by the symbol.

14. A finish that is angular in both directions on the surface is indicated by the symbol __?__ .

1. _____

2. _____

3. _____
   _____

4. _____

5. _____

6. _____
   _____

7. _____

8. _____

9. _____

10. _____

11. _____

12. _____

13. _____

14. _____

The following sketch may be found on a drawing to indicate the surface-finish requirements of a part.

Which symbols or figures in the sketch indicate the following specifications?

15. Roughness width

16. Surface finish (in microinches)

17. Waviness height

18. Direction of lay

15. _____

16. _____

17. _____

18. _____

SCORE _____

20

Measurement Review Test *(Section 5)*

PART 1

Place the name of each illustrated measuring tool in the proper space in the right-hand column.

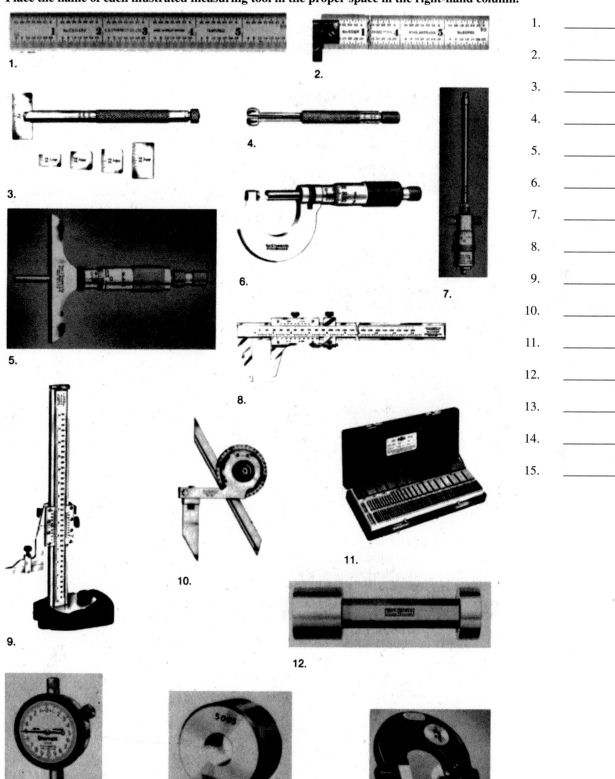

1.

2.

3.

4.

5.

6.

7.

8.

9.

10.

11.

12.

13.

14.

15.

1. _____

2. _____

3. _____

4. _____

5. _____

6. _____

7. _____

8. _____

9. _____

10. _____

11. _____

12. _____

13. _____

14. _____

15. _____

PART 2

Record the measuring-tool reading for each illustration in the proper space in the right-hand column.

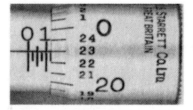

16.

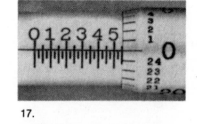

17.

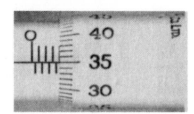

18.

19.

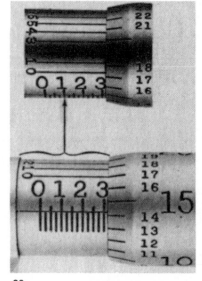

20.

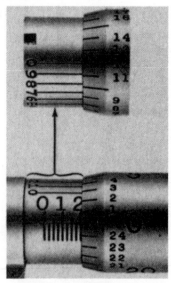

21.

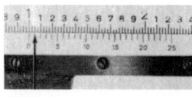

22.

23.

24.

25.

16. _____

17. _____

18. _____

19. _____

20. _____

21. _____

22. _____

23. _____

24. _____

25. _____

PART 3

Decide whether each statement is true or false and circle the letter in the right-hand column that indicates your choice.

26. Revolve the work in a lathe slowly when measuring with a caliper.   26.   T   F

27. A locking nut is not required on an inside micrometer since the spindle nut is a tight fit on the spindle thread.   27.   T   F

28. The rounded-type small hole gage is used for measuring shallow slots and holes.   28.   T   F

29. Gage blocks are manufactured only in 115- and 83-piece sets.   29.   T   F

30. Gage blocks, made of hardened alloy steel, are not affected by temperature change.   30.   T   F

PART 4

Place the correct word(s) in the blank space(s) provided at the right-hand side of the page that will make the sentence complete and true.

31. When taking a depth measurement, rotate the thimble of the depth micrometer in a ___?___ direction using the tip of one ___?___ .   31. _____

     _____

32. Gage blocks are accurate to size only at ___?___ °F.   32. _____

33. The laboratory or master set of gage blocks is accurate to within ___?___ of an inch.   33. _____

34. If the hole taper is not correct, the taper plug gage will ___?___ in the hole.   34. _____

35. Optical flats are used with a ___?___ light.   35. _____

36. Mechanical comparators may operate on the ___?___ and rack principle or use a system of ___?___ .   36. _____

     _____

PART 5

Select the most appropriate answer for each question and circle the letter in the right-hand column that indicates your choice.

37. On the micrometer spindle there are   37.   A   B   C   D
    (A) 1000 threads per inch   (C) 40 threads per inch
    (B) 25 threads per inch   (D) 100 threads per inch

38. Forty divisions on the sleeve of a micrometer are equal to   38.   A   B   C   D
    (A) .400 in.   (C) 1.000 in.
    (B) .040 in.   (D) .100 in.

39. To prevent rocking on an uneven surface, a surface plate is provided with   39.   A   B   C   D
    (A) three-point suspension   (C) four-point suspension
    (B) a wedge for leveling   (D) adjustable legs

40. One complete revolution of the micrometer thimble increases or decreases the distance between measuring faces by   40.   A   B   C   D
    (A) .025 in.   (C) .100 in.
    (B) .001 in.   (D) .0001 in.

41. A vernier micrometer has   41.   A   B   C   D
    (A) two scales on the thimble   (C) an indicating dial
    (B) two scales on the sleeve   (D) a sliding vernier bar

42. On a 50-division vernier, 50 divisions on the vernier scale are equal to   42.   A   B   C   D
    (A) 49 divisions on the bar   (C) 10 divisions on the bar
    (B) 50 divisions on the bar   (D) 100 divisions on the bar

43. It is a good habit to always expose the same face of the wear block to the work surface because   43.   A   B   C   D
    (A) any inaccuracy will be unilateral   (C) the accuracy life will be doubled
    (B) the wringing quality of the opposite face will be preserved   (D) known inaccuracies can be easily recognized

continued on next page

44. *Without the use of wear blocks,* the least number of gage blocks needed from an 83-piece set to make a 2.7005-in. buildup would be

(A) two   (C) four
(B) three   (D) five

44.   A   B   C   D

45. The universal bevel protractor has an accuracy of

(A) 5′   (C) 0.5°
(B) 5″   (D) 5°

45.   A   B   C   D

46. The 5-in. distance on a sine bar is measured

(A) by the length of the bar   (C) between the centers of the cylinders
(B) between the inside diameters of the cylinders   (D) between the outside diameters of the cylinders

46.   A   B   C   D

47. In theory, the sine bar becomes the

(A) hypotenuse of a right-angle triangle   (C) side adjacent of a right-angle triangle
(B) side opposite of a right-angle triangle   (D) right angle of a right-angle triangle

47.   A   B   C   D

48. When using a coordinate measuring system, one

(A) requires more skill   (C) must make several calculations before machining
(B) requires a good knowledge of mathematics   (D) reduces the chance of error

48.   A   B   C   D

SCORE   _____

50

Brad

TEST 18 Basic Layout Materials, *(Unit 19)* Tools, and Accessories

The first step in the production of a part is often the procedure of laying out the workpiece. Poor layout procedures can be costly since they often result in scrapped work or poorly fitting parts. A good layout can be produced only if the layout tools are used properly. The beginning machinist should therefore learn the correct care and use of these tools in order to produce accurate work.

Select the most appropriate answer for each question and circle the letter in the right-hand column that indicates your choice.

1. The most commonly used layout coating for metal is
   (A) chalk
   (B) copper sulphate
   (C) layout dye
   (D) torch bluing

   1.   A   B   Ⓒ   D

2. Copper sulphate should be used only on
   (A) ferrous metals
   (B) metals containing copper alloys
   (C) aluminum
   (D) all nonferrous metals

   2.   Ⓐ   B   C   D

3. Considered superior because they are harder, denser, and less porous, these surface plates are made of
   (A) cast iron
   (B) gray granite
   (C) pink granite
   (D) black granite

   3.   A   B   C   Ⓓ

4. Which of the following is not an advantage of granite surface plates over cast iron?
   (A) do not burr
   (B) do not rust
   (C) are magnetic
   (D) are cheaper

   4.   A   B   C   Ⓓ

5. The point of a scriber should be
   (A) hardened
   (B) hardened and tempered
   (C) ground to a 60° angle
   (D) ground to a 30° angle

   5.   A   Ⓑ   C   D

6. The marking points of layout tools are kept sharp by
   (A) grinding
   (B) honing
   (C) re-machining
   (D) filing

   6.   A   Ⓑ   C   D

7. A center punch has a point ground to an included angle of
   (A) 45°
   (B) 60°
   (C) 90°
   (D) 120°

   7.   A   B   Ⓒ   D

8. Trammels may be used for scribing circles when
   (A) a circle is laid out from a hole in the workpiece
   (B) the center has not been accurately laid out
   (C) the center-punch mark is too large
   (D) the surface is rough

   8.   Ⓐ   B   C   D

9. Trammels are used for scribing when
   (A) the circle to be scribed is beyond the range of dividers
   (B) all circles to be scribed are in excess of 6-in. diameter
   (C) dividers are not obtainable
   (D) greater accuracy is required

   9.   Ⓐ   B   C   D

10. Adjustable squares are used in layout work
    (A) when a high degree of accuracy is required
    (B) for general-purpose work
    (C) when an angle of greater than 90° is required
    (D) when an angle of less than 90° is required

    10.   A   Ⓑ   C   D

11. Which of the following layouts *cannot* be done with the steel rule and square head of a combination set?
    (A) a line parallel to a side
    (B) a 45° angle
    (C) a 90° angle
    (D) a 30° angle

    11.   A   B   C   Ⓓ

12. The square head and steel rule, used as a layout tool, is
    (A) very precise
    (B) fairly accurate
    (C) not a very useful tool
    (D) adjustable for angles

    12.   A   Ⓑ   C   D

continued on next page

13. On which of the following stock can a center *not* be laid out using a center head?   13.   A   B   ⟨C⟩   D
    (A) round   (C) hexagonal
    (B) square   (D) octagonal

14. When angles are being laid out with a bevel protractor, the best accuracy attainable is   14.   A   B   C   ⟨D⟩
    approximately
    (A) 5 minutes of arc   (C) 1° of arc
    (B) 1 minute of arc   (D) 30 minutes of arc

15. Which of the following is the most accurate method for scribing lines parallel to an edge?   15.   A   B   C   ⟨D⟩
    (A) a square head and steel rule   (C) a hermaphrodite caliper
    (B) a steel rule and scriber   (D) a surface gage

16. One of the disadvantages of a surface gage as a layout tool is that it   16.   A   B   ⟨C⟩   D
    (A) cannot be used on round stock   (C) cannot be adjusted to within .001 in.
    (B) has no provision for moving along an   of accuracy
        edge   (D) has no scriber adjustment

17. Lines may be drawn parallel to an edge with a surface gage by   17.   A   B   ⟨C⟩   D
    (A) moving the surface gage carefully   (C) depressing two pins in the base of the
        along the work parallel to the edge   surface gage
    (B) an accurate setting of the thumbscrew   (D) using the V-shaped groove in the base

18. Which of the following statements is *not true?* A surface gage may be used to   18.   A   B   C   ⟨D⟩
    (A) scribe lines parallel to a reference   (C) set up work in a vise
        surface   (D) accurately measure the height of a
    (B) scribe lines parallel to an edge   workpiece

19. Which of the following statements is *not true?* Angle plates may be   19.   A   B   ⟨C⟩   D
    (A) made of cast iron   (C) adjusted to any angle
    (B) used for layout work   (D) used for holding work for machining

20. Parallels are used in layout work to   20.   ⟨A⟩   B   C   D
    (A) raise a workpiece to the proper height   (C) provide proper measurement
    (B) set the top of an unfinished workpiece   (D) prevent damage to the surface plate
        parallel to a surface plate   by the workpiece

SCORE _____
                                                                                              20

TEST 19   Basic or Semiprecision   *(Unit 20)*
Layout

On all layout work, the layout should be made as simple as possible to save time and reduce the chance of error. Whenever possible, use only the basic layout tools unless the workpiece requires a precision layout.

Place the correct word(s) in the blank space(s) provided at the right-hand side of the page that will make the sentence complete and true.

1. When cutting stock, prior to layout, additional material should be allowed
   for __?__ the ends if required.

    1. _____

2. Angular layouts at corners are generally not shown on the print as angles but as the
   __?__ from the corner on __?__ adjacent sides.

    2. _____
      _____

3. Before starting any layout, remove all __?__ and __?__ the face of the workpiece.

    3. _____
      _____

4. All layout lines should be lightly __?__ -punched.

    4. _____

5. When no tolerances are shown on the print, a tolerance of ± __?__ ° is acceptable on
   an angle.

    5. _____

6. The adjustable __?__ is often used to lay out angles of 45°.

    6. _____

7. When a radius is being laid out on a corner, the center of the quadrant is located
   equidistant from two __?__ sides.

    7. _____

8. Place a __?__ under a short part to elevate, when clamping to an angle plate for layout.

    8. _____

9. All measurements for any location must be taken from the __?__ line or a __?__ edge.

    9. _____
      _____

10. When laying out flat stock a divider is used to scribe __?__ and __?__ .

    10. _____
      _____

11. To machine a hole concentric on a cored hole casting, fill the hole with a __?__
    piece and use a __?__ caliper to scribe four arcs, using the outside diameter
    as a reference surface.

    11. _____
      _____

12. When laying out a keyseat, use a __?__ rule or a __?__ gage to connect the circles.

    12. _____
      _____

13. When laying out a keyseat, use a __?__ punch to mark the layout and a __?__ punch to
    mark the circle centers.

    13. _____
      _____

SCORE _____

20

TEST 20   Precision Layout *(Unit 21)*

Precision layout is often necessary when accurate dimensions and hole positions are required on a workpiece. Since the accuracy of the workpiece is generally determined by the layout, it is important that care be taken when one is making a precision layout. The vernier height gage, sine bar, and gage blocks are often used for this purpose.

Place the correct word(s) in the blank space(s) provided at the right-hand side of the page that will make the sentence complete and true.

1. Hole locations and dimension lines are generally made from two machined adjacent edges called __?__ surfaces.

2. Measurements are made from the adjacent edges using X and Y __?__.

3. The vernier height gage may be used to mark off vertical distances to __?__ accuracy.

4. The vernier height gage consists of the __?__ , __?__ , and __?__ slide.

5. A __?__ is attached to the slide when making layouts.

6. Before making a layout, clean the surface of the __?__ table, the angle __?__, and the __?__ of the vernier height gage.

7. An __?__ scriber on the vernier height gage allows settings to be taken from the top of the surface plate.

8. If the zero on the vernier scale does not coincide with the zero on the beam when the scriber is touching the surface plate, __?__ the base of the height gage and the surface plate.

9. If the zeros still do not coincide, check the assembly of the __?__ and the vernier __?__ .

10. Holes may be accurately located by using the Coordinate Factors and __?__ Table.

11. Accurate hole and angular layout may be achieved by using a __?__ bar, __?__ blocks, and a __?__ height gage.

12. Angles and line lengths are calculated by means of __?__ tables.

13. After a layout has been made, it should be carefully __?__ before machining the workpiece.

1. _____

2. _____

3. _____

4. _____
   _____
   _____

5. _____

6. _____
   _____
   _____

7. _____

8. _____

9. _____
   _____

10. _____

11. _____
    _____
    _____

12. _____

13. _____

SCORE _____

20

Layout Tools and Procedures *(Section 6)*
Review Test

PART 1

1. Name the correct tools required to produce the marks or lines indicated on the drawing and place your answers in the right-hand column.

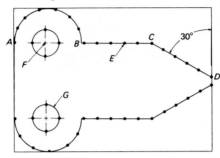

1. AB *scriber*

   BC *square*

   CD *bevel protract*

   E *prick punch*

   F *center punch*

   G *divider*

PART 2

Place the name of each illustrated layout tool or accessory in the proper space in the right-hand column.

2.

3.

4.

5.

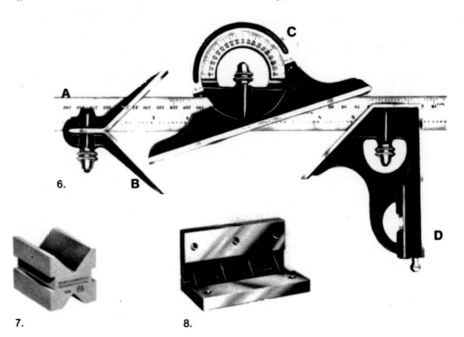

6.

7.

8.

2. *parallels*

3. *c-clamp*

4. *parallel clamp*

5. *trammell*

6. A *steel rule*

   B *center head*

   C *bevel protractor*

   D *square head*

7. *v-block*

8. *angle plate*

continued on next page

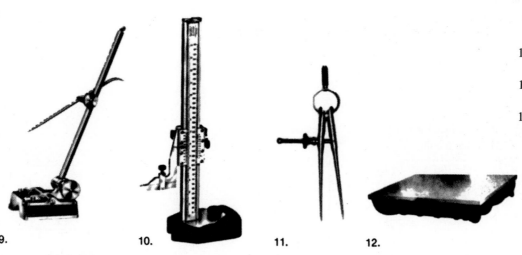

9.   Surface gage

10.   Venier-height

11.   divider

12.   Cast iron surface plate

9.   10.   11.   12.

PART 3

Select the most appropriate answer for each question and circle the letter in the right-hand column that indicates your choice.

13. The factor that determines the type of layout tools to be used is
    (A) availability of tools   (C) complexity of layout
    (B) degree of accuracy required   (D) defined by the part print
         13.   A   **B**   C   D

14. Copper sulphate should be used only on
    (A) ferrous metals   (C) aluminum
    (B) metals containing copper alloys   (D) nonferrous metals
         14.   **A**   B   C   D

15. For any layout to be accurate, it requires
    (A) bold, solid lines   (C) deep, wide lines
    (B) thick, clear lines   (D) fine, clear lines
         15.   A   B   C   **D**

16. Which of the following layouts *cannot* be done with a square head and steel rule?
    (A) a line parallel to an edge   (C) a 90° angle
    (B) a 45° angle   (D) a 30° angle
         16.   A   B   C   **D**

17. For laying out lines parallel to an edge, the most accurate layout tool is
    (A) a square head and rule   (C) a hermaphrodite caliper
    (B) a rule and scriber   (D) a surface gage
         17.   **A**   B   C   D

18. The maximum accuracy that can be achieved with an inch vernier height gage is
    (A) .010 in.   (C) .0001 in.
    (B) .001 in.   (D) .100 in.
         18.   A   **B**   C   D

19. The most accurate setting can be made on a height gage with the use of
    (A) gage blocks   (C) an adjustable vernier
    (B) an accurate surface plate   (D) an accurate angle plate
         19.   **A**   B   C   D

20. When a surface gage or vernier height gage is used for layout, which of the following statements is *incorrect?*
    (A) The scriber is at 45° to the surface.   (C) Clean the base of the gage.
    (B) Push, do not pull.   (D) Clean the surface plate.
         20.   A   **B**   C   D

21. Lines may be drawn parallel to an edge with a surface gage by
    (A) moving it carefully along the work   (C) holding to two base pins against the
           parallel to the edge   work edge
    (B) an accurate setting of the thumbscrew   (D) using the V-shaped groove in the base
         21.   **A**   B   C   D

22. When angles are being laid out with a bevel protractor, the best accuracy attainable is approximately
    (A) 5′ of a degree   (C) 1°
    (B) 1′ of a degree   (D) ½°
         22.   **A**   B   C   D

23. To achieve extreme accuracy in angular layout, use a
    (A) V-block and vernier height gage   (C) bevel protractor
    (B) universal bevel protractor   (D) sine bar and gage blocks

23. A (B) C D

24. The center of which of the following sections *cannot* be found with a center head and rule?
    (A) octagonal   (C) square
    (B) hexagonal   (D) circular

24. A (B) C D

25. Which of the following procedures would be correct to mark the center of a hole location?
    (A) Use a prick punch.   (C) Use a prick punch followed by a center punch.
    (B) Use a center punch.   (D) Use an automatic center punch.

25. A (B) C D

PART 4

Decide whether each statement is true or false and circle the letter in the right-hand column that indicates your choice.

26. If the part does not have to be too accurate, time should not be spent in making a precise layout.

26. (T) F

27. A cast-iron surface plate is better than a granite surface plate because it will not burr.

27. T (F)

28. The surface gage method of scribing a line parallel to a surface eliminates the error caused by a hand-held scriber.

28. (T) F

29. To permit a height gage to be used on a cylindrical surface, a round groove is machined into the base.

29. T (F)

30. To accurately place the prick punch on the layout line, hold the punch at 45° and place the point on the layout line.

30. T (F)

31. Hold the surface gage so that the scriber point is approximately 90° to the surface being laid out.

31. T (F)

32. In angular layout, accuracy of approximately 5 minutes may be achieved with a bevel protractor.

32. T (F)

33. Layout lines should be permanently marked with a center punch.

33. T (F)

34. The only difference between a center punch and a prick punch is the angle of the point.

34. (T) F

35. When scribing a line with a surface gage, always push the gage and scriber along the surface of the work.

35. T (F)

PART 5

Place the correct word(s) in the blank space(s) provided at the right-hand side of the page that will make the sentence complete and true.

36. The layout table least affected by temperature change is made of __?__ .

36. granite

37. Copper-sulphate layout solution should be used only on __?__ metals.

37. ferrous

38. The combination set consists of a rule, square, bevel protractor, and __?__ head.

38. center

39. A vernier height gage is capable of scribing lines to within __?__ of accuracy.

39. .001

40. A __?__ bevel protractor can be used to lay out angles to an accuracy of 5 minutes.

40. universal

41. Large circles or radii can be scribed with __?__ .

41. scriber

42. Round work can be held in a __?__ for layout or inspection purposes.

42. v-block

SCORE _____

50

TEST 21   Noncutting-Type Hand Tools *(Unit 22)*

The development of more accurate and versatile machine tools has greatly reduced the number of machine shop operations that still require the use of hand tools. However, hand tools are still important, especially when a machine tool is not available for a particular operation. It is important for the machinist to develop skills in the use of hand tools; such skills are generally acquired through practice over a period of years.

Select the most appropriate answer for each question and circle the letter in the right-hand column that indicates your choice.

1. To keep work from being damaged when it is held in a vise,   1. A **B** C D
   (A) limit the amount of pressure of the jaws   (C) use steel vise caps
   (B) use soft-metal vise caps   (D) hold the work so that the least damage will result

2. The most common hammer used by the machinist is a   2. A B C **D**
   (A) soft-faced hammer   (C) claw hammer
   (B) toolmaker's hammer   (D) ball-peen hammer

3. The large striking surface of a ball-peen hammer is the   3. A B **C** D
   (A) claw   (C) face
   (B) peen   (D) handle

4. The hammer head should be held tightly on the handle by means of a   4. A B **C** D
   (A) nail   (C) wedge
   (B) wood screw   (D) rivet

5. The blades of larger screwdrivers are usually   5. A **B** C D
   (A) round   (C) oval
   (B) square   (D) flat

6. Which of the following is *not* a common type of screwdriver?   6. A B **C** D
   (A) offset   (C) octagonal shank
   (B) Phillips   (D) stubby

7. A standard flat screwdriver will maintain a better grip in the slot if the blade is ground slightly   7. A B **C** D
   (A) tapered   (C) concave
   (B) convex   (D) rounded

8. The name of a wrench is derived from its   8. A B **C** D
   (A) use   (C) both A and B
   (B) shape   (D) neither A nor B

9. Which of the following statements does *not* apply to open-end wrenches?   9. **A** B C D
   (A) Openings are usually offset at 45°.   (C) They can be used in limited space by flopping them over.
   (B) Openings are usually offset at 15°.   (D) Double-end wrenches are of different sizes.

10. Combination pliers   10. **A** B C D
    (A) may be adjusted to grip larger workpieces   (C) are designed to reach into small places
    (B) make a good substitute for a wrench   (D) are used exclusively for wire cutting

Place the correct word(s) in the blank space(s) provided at the right-hand side of the page that will make the sentence complete and true.

11. Never use a hammer when your hands are __?__ .   11. _greasy_

12. If too small a screwdriver is used, both the __?__ in the screw and the __?__ of the blade will be damaged.   12. _slot_ _tip_

13. When regrinding a screwdriver, always grind an ___?___ amount off each side.

14. A box-end wrench has ___?___ precisely cut notches around the inside face.

15. Box-end wrenches are preferred over other wrenches because they cannot ___?___ if the proper size is used.

16. ___?___ cut files have two interesting rows of teeth.

17. Whenever possible, always ___?___ rather than ___?___ on a wrench.

18. Vise-grip pliers provide a high gripping power because of the adjustable ___?___ action.

13. _Equal_

14. _12_

15. _Slip_

16. _hex angle_

17. _pull_
    push

18. _lever_

SCORE _____

20

TEST 22   Hand-Type Cutting Tools (1) *(Unit 23)*

Although most metal cutting is done by machines more easily and quickly, it is often necessary to perform metal cutting and removal using hand-operated cutting tools. Skill in these operations can only be gained through experience on the job.

Place the correct word(s) in the blank space(s) provided at the right-hand side of the page that will make the sentence complete and true.

1. The three main parts of a hand hacksaw are the __?__ , __?__ , and __?__ .

    1. *frame*
       handle
       blade

2. Solid-frame hacksaws will accommodate blades only __?__ length.

    2. *Specific*

3. Hacksaw blades are made from high- __?__ steel.

    3. *speed*

4. Solid blades are __?__ throughout and are very __?__ .

    4. *hardened*
       brittle

5. Only the teeth on the __?__ blade are hardened.

    5. *flexible*

6. The __?__ is the most important factor to consider when selecting a hacksaw blade for a job.

    6. *pitch*

7. Always have at least __?__ saw teeth contacting the work when sawing.

    7. *two*

8. Mount the workpiece in the vise so that the cut will be about __?__ in. from the vise jaw.

    8. *.250*

9. When hacksawing, use about __?__ strokes per minute.

    9. *50*

10. When sawing, apply pressure only on the __?__ stroke.

    10. *forward*

11. A long angle lathe file is a __?__ -cut file.

    11. *single*

12. Never use a file as a __?__ or a hammer.

    12. *pry*

13. Applying __?__ to the face of the file will reduce the tendency for the file to become pinned.

    13. *Chalk*

14. When filing, never apply pressure on the __?__ stroke.

    14. *return*

15. __?__ filing is used to produce a smooth surface.

    15. *Draw*

16. The flatness of a piece of work should be checked occasionally with the edge of a __?__ .

    16. *file*

17. Never start a new saw blade in an __?__ cut.

    17. *Old*

SCORE _____

20

Bastard
Second
Smooth

TEST 23   Hand-Type Cutting Tools (2)   *(Unit 23)*

Select the most appropriate answer for each question and circle the letter in the right-hand column that indicates your choice.

1. After a work surface has been filed flat and smooth, the small scratches should be removed by
   - (A) buffing
   - (B) using abrasive cloth
   - (C) grinding
   - (D) milling

   1. A **B** C D

2. Aluminum files are used because they
   - (A) do not clog easily
   - (B) produce small scallops that cause the chip to break up easily
   - (C) clear the chips easily
   - (D) all of the previous

   2. A B C **D**

3. Which of the following statements is *not true* regarding the angle of teeth on a lathe file?
   - (A) It tends to clear the teeth.
   - (B) It helps eliminate chatter.
   - (C) It reduces the possibility of tearing the metal.
   - (D) It makes a more accurate cut.

   3. A B C **D**

4. Which of the following files is *not* a precision file?
   - (A) Swiss pattern
   - (B) needle
   - (C) microfinish
   - (D) riffler

   4. A B **C** D

5. Which of the following statements does *not* apply to ground burrs?
   - (A) They have better chip clearance than rotary files.
   - (B) They cut aluminum more efficiently than rotary files.
   - (C) They cut steel more efficiently than rotary files.
   - (D) They are ground from a master burr.

   5. A B **C** D

Decide whether each statement is true or false and circle the letter in the right-hand column that indicates your choice.

6. Thin metal may be held for sawing between two pieces of wood.   6. **T** F

7. The teeth on a hacksaw blade should point away from the handle.   7. **T** F

8. Pressure should only be applied to the file on the forward stroke.   8. **T** F

9. The most common degrees of file coarseness used in a machine shop are the bastard, second cut, and smooth.   9. **T** F

10. Aluminum files may be used efficiently on white cast iron.   10. T **F**

11. A file should be cleaned by knocking its edges on the vise.   11. T **F**

12. Oil should be applied when using a new file to keep the teeth sharp.   12. T **F**

13. Excessive pressure on a new file will shorten the life of the file.   13. **T** F

14. Die sinker rifflers have straight ends to scrape the bottom of the die cavity.   14. T **F**

15. Rotary files are designed for use on work that rotates.   15. T **F**

16. Ground burrs are more efficient than rotary files for aluminum.   16. **T** F

17. Ground burrs have a greater chip clearance than rotary files.   17. **T** F

18. Carbide burrs last 1000 times longer than high-speed burrs.   18. T **F**

19. Carbide burrs can be used on hard and soft materials.   19. **T** F

20. Scraping can produce a truer surface than milling.   20. **T** F

SCORE _____

20

TEST 24   Thread Cutting *(Unit 24)*

Threads have many applications in the machine tool industry. They may be used to hold parts together, for transmission of motion, as in a lead screw or in a screw-type jack, or for measurement, as in a micrometer. Although the operation of thread cutting by hand is quite simple, care must be taken not to break the tap or die and possibly ruin the work.

Place the correct word(s) in the blank space(s) provided at the right-hand side of the page that will make the sentence complete and true.

1. A tap is used to cut an __?__ thread.

2. Flutes are cut in a tap to form __?__ edges, to provide clearance for __?__, and to admit cutting __?__ .

3. Identify each part of the tap nomenclature: .500 in. __?__ , 13 __?__ , UNC __?__.

4. The three hand taps in a set are called __?__, __?__, and __?__.

5. The tap drill leaves enough material in the hole to produce __?__ percent of a full thread.

6. The tap-drill size TDS is equal to $\frac{?}{} - \frac{1}{?}$.

7. When checking the tap for squareness, always check it from two positions __?__ ° from each other.

8. No cutting fluid is required when tapping __?__ or cast iron.

9. Before using a bottoming tap, remove all __?__ from the hole.

10. For removing a broken tap, a tap __?__ might be successful.

11. If a broken tap is made of __?__ steel, it may be possible to drill it out after heating it and cooling it __?__ .

12. A __?__ die is used for chasing or recutting damaged threads.

13. An __?__ split die permits an adjustment over or under the standard thread size.

14. When cutting a thread with a die, always apply equal __?__ to each handle of the die stock.

15. Turn the die forward one-quarter turn and then reverse it one- __?__ turn to __?__ the chip.

16. Before threading a piece of round stock with a die, always __?__ the end.

1. _____

2. _____

_____

_____

3. _____

_____

_____

4. _____

_____

_____

5. _____

6. _____

7. _____

8. _____

9. _____

10. _____

11. _____

12. _____

13. _____

14. _____

15. _____

16. _____

SCORE _____

25

TEST 25   Reaming and Broaching *(Unit 25)*

Reaming, although preferably done on a machine tool, may be performed by hand, especially when parts are being assembled and the use of a machine such as a drill press is impossible.

Broaching is used to produce specially shaped holes in metal, and hand (nonmachine) broaching is generally done using an arbor press.

Select the most appropriate answer for each question and circle the letter in the right-hand column that indicates your choice.

1. Which of the following statements is *not true* in regard to solid reamers?
   (A) They are not adjustable.
   (B) The end is ground straight.
   (C) They have straight flutes.
   (D) They have helical flutes.

   1.   A   B   C   D

2. Hand reamers are designed to remove no more than
   (A) .005 to .010 in.
   (B) .001 to .002 in.
   (C) .010 to .015 in.
   (D) .015 to .020 in.

   2.   A   B   C   D

3. Expansion hand reamers of up to .50-in. diameter permit an adjustment of approximately
   (A) .001 in.
   (B) .003 in.
   (C) .0003 in.
   (D) .006 in.

   3.   A   B   C   D

4. Expansion hand reamers are adjusted to size by
   (A) a threaded plug
   (B) an adjustable wedge
   (C) a setscrew
   (D) two adjusting nuts

   4.   A   B   C   D

5. Adjustable hand reamers are adjusted to size by
   (A) a threaded plug
   (B) an adjustable wedge
   (C) a setscrew
   (D) two adjusting nuts

   5.   A   B   C   D

6. Which of the following statements is *not true* regarding tapered reamers?
   (A) They may have straight flutes.
   (B) They are adjustable.
   (C) They may have spiral flutes.
   (D) The chips must be cleared frequently.

   6.   A   B   C   D

7. A roughing taper reamer
   (A) has widely spaced flutes
   (B) has nicks ground at intervals along teeth
   (C) has narrowly spaced flutes
   (D) must be revolved counterclockwise

   7.   A   B   C   D

8. Which of the following statements is *not true* regarding a finishing taper reamer?
   (A) It has straight flutes.
   (B) It has right-hand spiral flutes.
   (C) It has left-hand spiral flutes.
   (D) It does not clear the chips readily.

   8.   A   B   C   D

9. Which of the following statements is *not true* regarding broaching?
   (A) It can produce irregularly shaped holes
   (B) Hole size and shape are not consistent.
   (C) It produces forms quickly.
   (D) Roughing and finishing cuts are performed in one operation.

   9.   A   B   C   D

10. Which of the following statements is *not true* regarding the reaming of a hole in steel?
    (A) Always turn the reamer clockwise.
    (B) Use cutting fluid.
    (C) Always turn the reamer counterclockwise.
    (D) Clear the chips frequently.

    10.   A   B   C   D

SCORE _____

10

TEST 26   Surface Finishes *(Unit 26)*

Surface treatment is used on many metals to resist wear, stop electrolytic decomposition, and prevent corrosive wear due to exposure to weather. The surface of metals is treated by a number of different methods, depending upon the metal being treated.

Place the correct word(s) in the blank space(s) provided in the right-hand side of the page that will make the sentence complete and accurate.

1. A cold-working process that sizes, __?__ , and work hardens metal surfaces by pressure of hardened rolls is called __?__ .

2. The result of the process is a mirror-like finish, tough __?-?__ , and wear- and __?__ resistant surface.

3. Electropolishing is often referred to as a __?-?__ process.

4. Honing is an __?__ finishing operation that removes a small amount of stock to improve a part, __?__ , and surface finish.

5. There are two common types of honing operations, conventional and __?-?__ .

6. Honing speeds are __?__ than grinding speeds.

7. When selecting a hone, the type and __?__ of the work material and the __?__ of material removed must be considered.

8. Tumbling is used to clean, __?__ , and remove __?__ from metal parts.

9. Tumbling media can consist of aluminum oxide, __?__ , stone chips, __?__ , crushed corncob, or steel balls.

10. Black-oxide coatings are a __?__ conversion process, produced by the __?__ of iron to the oxidizing salts.

1. _____

_____

2. _____ -

_____

_____

3. _____ -

4. _____

_____

5. _____

6. _____ -

7. _____

_____

8. _____

_____

9. _____

_____

10. _____

_____

SCORE _____

20

Hand Tools and Benchwork *(Section 7)*
Review Test

PART 1

Place the name of each illustrated tool in the proper space in the right-hand column.

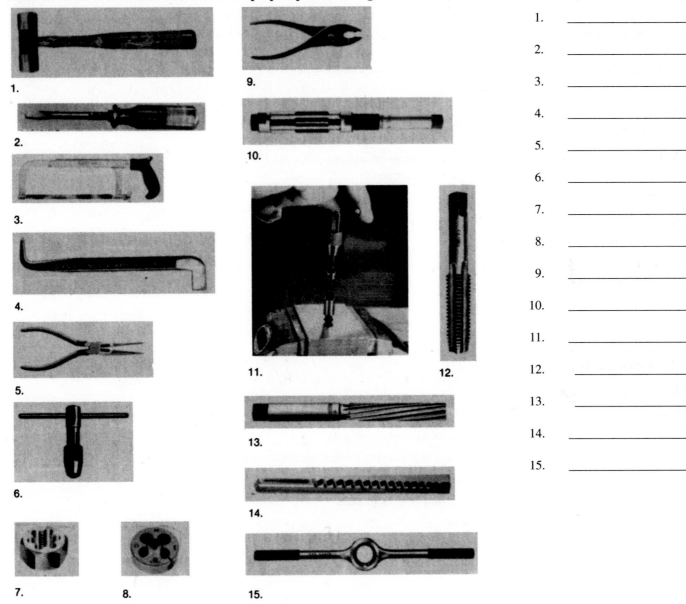

1.

2.

3.

4.

5.

6.

7.   8.

9.

10.

11.   12.

13.

14.

15.

1. _____

2. _____

3. _____

4. _____

5. _____

6. _____

7. _____

8. _____

9. _____

10. _____

11. _____

12. _____

13. _____

14. _____

15. _____

PART 2

Select the most appropriate answer for each question and circle the letter in the right-hand column that indicates your choice.

16. The size of a vise is determined by the
    - (A) jaw opening
    - (B) length of vise jaw
    - (C) width of jaws
    - (D) thickness of jaws

    16. A   B   C   D

17. To avoid damaging work held in a vise,
    - (A) limit the amount of pressure on the jaws
    - (B) use soft-metal vise caps
    - (C) use steel vise caps
    - (D) hold the work in such a manner that the least damage will result

    17. A   B   C   D

continued on next page

18. The most common hammer used by a machinist is a
    (A) soft-faced hammer
    (B) toolmaker's hammer
    (C) claw hammer
    (D) ball-peen hammer

    18.　A　　B　　C　　D

19. When you are regrinding the blade of a screwdriver, which of the following practices is *incorrect?*
    (A) Remove a minimum amount of metal.
    (B) Do not quench in water; overheating improves the temper.
    (C) Regrind the tip slightly concave, not straight.
    (D) Grind an equal amount off each side.

    19.　A　　B　　C　　D

20. To cut thin material with a hand hacksaw, use a blade with
    (A) a coarse pitch
    (B) maximum chip clearance
    (C) a fine pitch
    (D) minimum chip clearance

    20.　A　　B　　C　　D

21. Which of the following practices does *not* apply when sawing by hand?
    (A) Use about 120 strokes per minute.
    (B) Ensure that the teeth face away from the handle.
    (C) See that two or more teeth are on the section to be cut.
    (D) Apply pressure on the forward stroke.

    21.　A　　B　　C　　D

22. Which of the following terms does *not* define a degree of coarseness of a file?
    (A) double cut
    (B) second cut
    (C) rough
    (D) smooth

    22.　A　　B　　C　　D

23. Files are usually made of
    (A) high-carbon steel
    (B) low-carbon steel
    (C) medium-carbon steel
    (D) high-speed steel

    23.　A　　B　　C　　D

24. To clean the cuttings from a file,
    (A) use an air hose
    (B) tap it against a metal object
    (C) use a sharp pointed instrument
    (D) use a file card

    24.　A　　B　　C　　D

25. A smooth flat surface is produced by
    (A) cross filing
    (B) hand filing
    (C) draw filing
    (D) finish filing

    25.　A　　B　　C　　D

PART 3

Decide whether each statement is true or false and circle the letter in the right-hand column that indicates your choice.

26. The size of a vise is determined by the size of the work it will hold.　　26.　T　　F

27. To keep work from being damaged when it is held in a vise, use hardened steel vise caps.　　27.　T　　F

28. Hacksaw blades are made of medium-carbon steel to prevent them from breaking easily.　　28.　T　　F

29. The teeth of a hacksaw blade should point away from the handle.　　29.　T　　F

30. Phillips-type screwdrivers have a square tip.　　30.　T　　F

31. Always pull rather than push a wrench when tightening or loosening a nut.　　31.　T　　F

32. A plug tap is tapered for approximately six threads for easier starting.　　32.　T　　F

33. The solid die is used for chasing and recutting damaged threads.　　33.　T　　F

34. An expansion hand reamer has an adjustment range of .030 in.　　34.　T　　F

35. When removing a reamer from a hole, turn it in a counterclockwise direction.　　35.　T　　F

PART 4

Place the correct word(s) in the blank space(s) provided at the right-hand side of the page that will make the sentence complete and true.

36. A vise equipped with a __?__ base can be swung to any circular position.　　36. _____

37. Only the teeth of flexible hacksaw blades are __?__ .　　37. _____

38. Pressure should be applied only on the __?__ stroke of the file.　　38. _____

39. Apply ___?___ to the face of a file to lessen the tendency for it to become "pinned."

40. Pliers should never be used as a ___?___ .

41. Flutes are cut in a tap to form the ___?___ edge.

42. A plug tap has the first ___?___ threads tapered for easy starting.

43. The tap drill leaves enough material in the hole to produce ___?___ percent of a full thread.

44. Before using the bottoming tap, remove all ___?___ from the hole.

45. If a broken tap is made of ___?___ steel, it may be possible to drill it out after heating and cooling it slowly.

46. No cutting fluid is required when tapping ___?___ or cast iron.

47. The cutting end of a hand reamer is ground to a slight taper for a distance equal to the ___?___ of the reamer.

48. No more than ___?___ should be left in a hole for hand reaming.

49. Helical fluted hand reamers tend to reduce ___?___ because of their shearing action.

50. Straight fluted hand reamers should not be used on work that has a ___?___ .

39. _____

40. _____

41. _____

42. _____

43. _____

44. _____

45. _____

46. _____

47. _____

48. _____

49. _____

50. _____

SCORE _____

50

TEST 27   Machinability of Metals *(Unit 28)*

Machinability refers to the ease or difficulty with which any material can be machined (turned, milled, etc.). This is generally measured by the tool life in minutes or the material removal rate in relation to the cutting speed and the depth of cut. Hard materials generally have low machinability, while softer materials have high machinability.

Place the correct word(s) in the blank space(s) provided at the right-hand side of the page that will make the sentence complete and true.

1. The machinability of a metal is affected by its microstructure and will vary if the metal has been __?__ , __?__ , or ___?-?___ .

   1. _____
   _____
   _____

2. High carbon steel contains a greater amount of __?__ than does low carbon steel.

   2. _____

3. The machining qualities of alloy steels can generally be improved by the addition of sulfur and __?__ or sulfur and __?__ .

   3. _____
   _____

4. Stainless steels are generally difficult to machine because of their ___?-?___ qualities.

   4. _____

5. __?__ and __?__ aluminum alloys are generally easier to machine than annealed alloys.

   5. _____
   _____

6. In any metal removal process, heat is created by __?__ __?__ and __?__ .

   6. _____
   _____
   _____

7. The maximum temperatures created during the cutting action will affect the __?__ __?__ , __?__ __?__ , __?__ __?__ , and accuracy of the workpiece.

   7. _____
   _____
   _____
   _____
   _____
   _____

8. A good supply of __?__ __?__ will help to reduce the friction at the ___?-?___ interface and maintain efficient cutting temperatures.

   8. _____
   _____
   _____

9. Four ways in which cutting fluids assist the machining of metals are __?__ , __?__ , __?__ , and __?__ .

   9. _____
   _____
   _____
   _____

   SCORE _____
   20

TEST 28   Cutting Tools and Operating Conditions   *(Units 29, 30)*

Two of the most important factors that must be considered for any machining process are the type and shape of the cutting tool. These two factors will determine the rate of material removal, the life and efficiency of the cutting tool, and the productivity of the machining operation.

Select the most appropriate answer for each question and circle the letter in the right-hand column that indicates your choice.

1. Which of the following statements is *not true* regarding a lathe cutting tool?
   (A) It should be hard.
   (B) It should be wear resistant.
   (C) It should be capable of withstanding high temperatures.
   (D) It should be fairly brittle.

   1.   A   B   C   D

2. High-speed steel cutting tools are generally made of two base materials to which other alloys are added. These are
   (A) chromium and vanadium
   (B) chromium and tungsten
   (C) tungsten and molybdenum
   (D) tungsten and vanadium

   2.   A   B   C   D

3. Which of the following statements is *not* a characteristic of cast-alloy cutting tools?
   (A) excellent red hardness
   (B) weaker than high-speed steel cutting tools
   (C) more brittle than high-speed steel
   (D) suited for deep interrupted cuts

   3.   A   B   C   D

4. Red hardness, as applied to a cutting tool, means
   (A) the cutting tool will maintain a sharp edge at red heat
   (B) the work speed is too high if the cutting tool becomes red
   (C) the cutting edge will break down if the cutting tool becomes red-hot
   (D) the hardness will be maintained only if the cutting tool is red-hot

   4.   A   B   C   D

5. When more red hardness is required, the cutting tool should contain more
   (A) vanadium
   (B) cobalt
   (C) tungsten
   (D) molybdenum

   5.   A   B   C   D

6. Which of the following properties is *incorrect* in regard to carbide cutting tools?
   (A) high toughness
   (B) high hardness
   (C) red hardness
   (D) a cutting speed of 3 to 4 times faster than high-speed steel

   6.   A   B   C   D

7. The side-cutting edge angle of a lathe cutting tool should be from
   (A) 15 to 30°
   (B) 10 to 20°
   (C) 5 to 10°
   (D) 22 to 25°

   7.   A   B   C   D

8. If the side-cutting edge angle is more than 30°, the tool will tend to
   (A) produce a fine finish
   (B) chatter
   (C) dull quickly
   (D) break

   8.   A   B   C   D

9. For roughing cuts, the end-cutting edge angle should be
   (A) 1 to 5°
   (B) 5 to 10°
   (C) 5 to 15°
   (D) 15 to 30°

   9.   A   B   C   D

10. For general-purpose use, the end-cutting edge angle should be
    (A) 1 to 5°
    (B) 5 to 10°
    (C) 5 to 15°
    (D) 15 to 30°

    10.   A   B   C   D

11. A larger end-cutting angle is satisfactory for
    (A) facing
    (B) cutting close to the chuck
    (C) turning shoulders
    (D) all of these

    11.   A   B   C   D

12. The side-relief, or side-clearance, angle should be
    (A) 2 to 5°
    (B) 6 to 10°
    (C) 12 to 15°
    (D) 15 to 30°

    12.   A   B   C   D

continued on next page

13. The end-relief angle for general-purpose tools should be

    (A) 5 to 10°   (C) 15 to 20°

    (B) 10 to 15°   (D) 20 to 25°

13.   A   B   C   D

14. The end-relief angle is measured when

    (A) the shank of the cutting tool is placed   (C) the cutting tool is set up in the lathe

        on a flat surface   (D) the cutting tool is mounted in the toolholder,

    (B) the cutting tool is held in the toolholder   which is placed on a flat surface

14.   A   B   C   D

15. The side-rake angle for general-purpose cutting tools should be about

    (A) 6°   (C) 14°

    (B) 10°   (D) 18°

15.   A   B   C   D

16. The angle of keenness is formed by grinding

    (A) side rake and side clearance   (C) side rake and front clearance

    (B) back rake and side clearance   (D) back rake and front rake

16.   A   B   C   D

Place the correct word(s) in the blank space(s) provided at the right-hand side of the page that will make the sentence complete and true.

17. The three operating conditions that affect the metal-removal rate when machining are ___?___ , ___?___ , and ___?___ .

17. _____

     _____

     _____

18. The operating condition that causes the least reduction in tool life is the ___?___ .

18. _____

19. The best cutting speed for any job must be one that balances the ___?-?___ rate and the ___?-?___ life.

19. _____ -

     _____ -

20. Two machine tool factors that can affect the production rate are ___?___ and ___?___ .

20. _____

     _____

21. Any cutting tool's ability to remove stock is governed by the number of times it must be ___?___ or replaced in order to produce accurate work and a good surface finish.

21. _____

                      SCORE _____

                                   25

TEST 29   Carbide, Ceramic, and *(Units 31, 32)* Diamond Cutting Tools

In order to remain competitive, any industry must look toward faster and more efficient ways to manufacture products. This search led to the development of cutting tools that would last longer and be capable of higher cutting speeds. As a result high-speed cutting tools have generally been replaced with carbide, ceramic, and diamond cutting tools, depending on the job requirements.

Select the most appropriate answer for each question and circle the letter in the right-hand column that indicates your choice.

1. The process used to produce cemented carbide is
   (A) electrolytic deposition   (C) crushing
   (B) molding   (D) powder metallurgy

   1.   A   B   C   D

2. Cobalt is used in the manufacture of cemented-carbide cutting tools to
   (A) reduce brittleness   (C) act as a binder
   (B) reduce cratering   (D) increase hardness

   2.   A   B   C   D

3. The carbide powders and cobalt are mixed with
   (A) oil   (C) kerosene
   (B) water   (D) alcohol

   3.   A   B   C   D

4. Which of the following statements does *not* apply to presintering?
   (A) Blanks have the consistency of chalk.   (C) It is carried out in a nitrogen atmosphere.
   (B) It is carried out at 1500°F.   (D) Compacts shrink about 40 percent.

   4.   A   B   C   D

5. The strongest and most wear-resistant grades of carbide contain tungsten carbide and cobalt plus
   (A) tantalum carbide   (C) tantalum and titanium carbides
   (B) titanium carbide   (D) no other carbides

   5.   A   B   C   D

6. The addition of titanium carbide increases
   (A) resistance to cratering   (C) wear resistance
   (B) toughness   (D) brittleness

   6.   A   B   C   D

7. Carbide inserts will give longer life and a freer chip flow if they are coated with
   (A) tungsten nitride   (C) zirconium oxide
   (B) titanium nitride   (D) aluminum oxide

   7.   A   B   C   D

8. Front or end relief (clearance) on a carbide cutting tool
   (A) may be positive or negative (C) is more for cast iron than machine steel
   (B) is the same for all metals   (D) will cause rapid tool failure if too great

   8.   A   B   C   D

9. The nose radius on a carbide cutting tool
   (A) should be about twice the amount of   (C) does not affect the surface finish of the work
       feed per revolutions per minute   (D) may cause chatter if too small
   (B) should be as large as possible

   9.   A   B   C   D

10. The amount of side rake is affected by three factors. Which of the following is *not* a factor?
    (A) feed per revolution   (C) surface feet per minute
    (B) type and grade of cutting tool   (D) type of material being cut

    10.   A   B   C   D

11. Various charts list the recommended angles for carbide cutting tools. The machinist should
    (A) adhere to these   (C) change the feed
    (B) change these slightly to suit various   (D) change the spindle speed
        conditions

    11.   A   B   C   D

12. The lathe center used with carbide cutting tools should be a
    (A) carbide center   (C) revolving dead center
    (B) live center   (D) dead center

    12.   A   B   C   D

continued on next page

13. A rocker-type carbide toolholder should have   13. A B C D
   (A) 5° positive rake   (C) 10° positive rake
   (B) 10° negative rake   (D) zero rake

14. Which of the following statements is *not* correct when a carbide cutting tool is used in a   14. A B C D
   standard lathe toolpost?
   (A) Shim the tool to the correct height.   (C) Set the tool to 5° back rake.
   (B) Invert the rocker base.   (D) Remove the rocker.

15. When machining with carbide cutting tools, apply cutting fluid   15. A B C D
   (A) constantly under pressure   (C) when the cutting tool overheats
   (B) intermittently   (D) only on cast iron

16. Which of the following statements is *not* correct with regard to honing a carbide lathe   16. A B C D
   tool for machining steel?
   (A) A 320-grit silicon-carbide stone should (C) A 45° chamfer should be honed on the
       be used.   cutting edge.
   (B) It should be honed to a sharp, keen   (D) Honing is very important to the life of
       edge.   the cutting tool.

17. Which of the following grinding wheels is recommended for rough grinding a carbide   17. A B C D
   cutting tool?
   (A) borazon   (C) diamond
   (B) aluminum oxide   (D) silicon carbide

18. Which of the following statements is *not true* regarding carbide cutting tools?   18. A B C D
   (A) They should be used only on large,   (C) The lathe should have sufficient power
       heavy-duty lathes.   to maintain a constant cutting speed.
   (B) The lathe should have hardened gears. (D) The lathe should be rigid.

19. Ceramic cutting tools are considered   19. A B C D
   (A) a supplement to carbide tools   (C) a total replacement for carbide tools
   (B) more suitable than carbides on less   (D) not suitable for castings because of the
       rigid machines   hard scale and sand

20. A newer and stronger ceramic cutting tool has been developed by mixing   20. A B C D
   (A) aluminum oxide and zirconium oxide (C) silicon carbide and zirconium oxide
   (B) aluminum oxide and magnesium oxide (D) silicon carbide and titanium oxide

21. Which of the following statements is *not* correct with regard to ceramic cutting tools?   21. A B C D
   (A) Negative rake inserts give the best   (C) Cutting fluids must always be used.
       results.   (D) The machinist must be capable of
   (B) A large cutting-edge angle reduces the   maintaining high speeds.
       tendency to chip.

22. If necessary, a ceramic tool may be ground with a   22. A B C D
   (A) resinoid-bonded diamond wheel   (C) vitrified-bonded diamond wheel
   (B) resinoid-bonded silicon-carbide wheel (D) vitrified-bonded silicon-carbide wheel

23. Diamond-tipped cutting tools are used to machine   23. A B C D
   (A) ferrous materials   (C) hardened steel
   (B) nonferrous materials   (D) alloy-steel castings

24. One of the main advantages of diamond cutting tools is that they   24. A B C D
   (A) can withstand interrupted cuts in hard (C) can take heavy cuts
       metal   (D) can operate at high speeds
   (B) are shock-resistant

25. Under which of the following conditions does a diamond cutting tool perform more   25. A B C D
   efficiently?
   (A) shallow cut and coarse feed   (C) low speed and fine feed
   (B) shallow cut and fine feed   (D) high speed and coarse feed

SCORE _____

25

TEST 30   Polycrystalline Cutting Tools *(Unit 33)*

The manufacture of polycrystalline cutting tool blanks was a major step in the production of new and more efficient cutting tools. Superabrasive high-efficiency cutting tools have a long-wearing superior cutting edge that has excellent abrasion resistance. They are capable of operating at higher cutting speeds, take deeper cuts, and can machine hard, abrasive materials and alloys.

Place the correct word(s) in the blank space(s) provided at the right-hand side of the page that will make the sentence complete and true.

1. The two types of polycrystalline cutting tools are __?__ and __?__ .   1. _____

_____

2. Polycrystalline tool blanks have a layer of superabrasive material fused to a __?__ -   2. _____ -
__?__ substrate.

3. The four main properties of polycrystalline cubic boron nitride are   3. _____
__?__ , __?__ , __?__ , and __?__ .

_____

_____

_____

4. The four general categories of metals that are cut most cost-efficiently with PCBN tools are   4. _____
__?__ , __?__ , __?__ , and __?__ .

_____

_____

_____

5. Four main advantages that PCBN tools offer industry are __?__ , __?__ , __?__ ,   5. _____
and __?__ .

_____

_____

_____

6. PCBN tools are used to machine __?__ materials while PCD tools are used to machine   6. _____
__?__ materials.

_____

7. The three microstructures in which polycrystalline diamond tools are available are   7. _____
__?__ , __?__ , and __?__ .

_____

_____

SCORE _____

20

TEST 31   Cutting Fluids *(Unit 34)*

Cutting fluids are essential in most metal-cutting operations to reduce the heat and friction created at the chip-tool interface. Excessive heat could cause metal to adhere to the tool's cutting edge, cause the tool to break down quickly, produce a poor surface finish, and produce inaccurate work.

Place the correct word(s) in the blank space(s) provided at the right-hand side of the page that will make the sentence complete and accurate.

1. The four economic advantages that cutting fluids offer are __?__, __?__, __?__, and __?__.   1. _____

   _____

   _____

   _____

2. The three places where the heat generated at the chip-tool interface can find its way into the tool   2. _____
   are the __?__, __?__, and __?__.

   _____

   _____

3. The two most important qualities that a cutting fluid must possess are __?__ and __?__.   3. _____

   _____

4. Good cutting fluids should also __?__ cutting tool life, resist __?__, and provide   4. _____
   __?__ control.

   _____

   _____

5. The three main categories of cutting fluids are __?__, __?__, and __?__.   5. _____

   _____

   _____

6. The two sources of heat generated during a cutting action are the __?__ __?__ of the   6. _____
   metal ahead of the cutting tool and the __?__ of the chip sliding over the tool face.

   _____

   _____

7. If a built-up edge is allowed to form on a cutting tool edge, it will result in __?__ surface   7. _____
   finish, excessive __?__ wear, and __?__ of the tool face.

   _____

   _____

   SCORE   _____

   20

PART 4

Place the correct word(s) in the blank space(s) provided at the right-hand side of the page that will make the sentence complete and true.

25. Polycrystalline tool blanks have a layer of superabrasive material fused to a ___?-?___ substrate.

25. _____

26. The four main properties of polycrystalline cubic boron nitride are ___?___ , ___?___ , ___?___ , and ___?___ .

26. _____
_____
_____
_____

27. PCBN tools are used to machine ___?___ materials while PCD tools are used to machine ___?___ materials.

27. _____
_____

28. The three microstructures in which polycrystalline diamond tools are available are ___?___ , ___?___ , and ___?___ .

28. _____
_____
_____

PART 5

29. The three places where the heat generated at the chip-tool interface can find its way into the tool are the ___?___ , ___?___ , and ___?___ .

29. _____
_____
_____

30. The two most important qualities that a cutting fluid must possess are ___?___ and ___?___ .

30. _____
_____

31. The two sources of heat generated during a cutting action are the ___?___ ___?___ of the metal ahead of the cutting tool and the ___?___ of the chip sliding over the tool face.

31. _____
_____
_____

32. If a built-up edge is allowed to form on a cutting tool edge, it will result in ___?___ surface finish, excessive ___?___ wear, and ___?___ of the tool face.

32. _____
_____
_____

SCORE _____

50

TEST 32   Cutoff Saws   *(Unit 35)*

Cutoff saws are used extensively in machine shops to cut work to the rough length from bar stock material. The horizontal bandsaw is probably the most popular saw used for this purpose.

Place the correct word(s) in the blank space(s) provided at the right-hand side of the page that will make the sentence complete and true.

1. Bandsaw blades are manufactured in varying degrees of coarseness from __?__ to __?__ pitch.

   1. _____

      _____

2. A __?__ -pitch blade is used for cutting large sections.

   2. _____

3. A __?__ -pitch blade is recommended for general-purpose work.

   3. _____

4. Bandsaw blades have only the __?__ hardened.

   4. _____

5. Too __?__ a blade speed or __?__ feeding pressure will dull the saw teeth quickly.

   5. _____

      _____

6. When installing a blade, be sure that the teeth point in the direction of the __?__ .

   6. _____

7. Before setting the work in the saw vise, be sure that the solid jaw is __?__ .

   7. _____

8. When several pieces of the same length are required, the __?__ gage should be used.

   8. _____

9. Long pieces of work should be __?__ with a floor stand.

   9. _____

10. When holding short work in a vise, always place a short piece of the __?__ thickness in the __?__ end of the vise.

    10. _____

        _____

11. Before taking the cut, adjust the roller guide __?__ so that they just __?__ the work.

    11. _____

        _____

12. Always allow __?__ in. extra length for each inch of thickness of the material to allow for the saw __?__ .

    12. _____

        _____

13. Before taking a cut, always __?__ the length of the material and see that the vise is __?__ .

    13. _____

        _____

14. When sawing material, always be sure that at least __?__ saw teeth are touching the work.

    14. _____

    SCORE _____

    20

TEST 33   Contour Bandsaw (*Unit 36*)

Since its introduction in the early 1930s, the contour bandsaw has found wide acceptance in the metalworking industries. Although bandsawing is not a precision metal-cutting operation, it has an advantage over other metal-removal operations in the saving of time and material.

Select the most appropriate answer for each question and circle the letter in the right-hand column that indicates your choice.

1. Compound angles can be cut by
   (A) clamping the work on an angle   (C) both A and B
   (B) tilting the table   (D) neither A nor B

   1.   A   B   C   D

2. A bandsaw blade, when properly selected, will hold its sharpness because
   (A) of the hardness of the saw teeth   (C) wear is distributed over many teeth
   (B) less horsepower is required to drive   (D) of the tooth set
   the narrow blade

   2.   A   B   C   D

3. Which of the following statements is *incorrect?*
   (A) Machining geometry is not restricted.   (C) Cutting force holds work on the table.
   (B) Only simple fixturing is required.   (D) Material removed is reduced to chips.

   3.   A   B   C   D

4. The saw blade that will produce the best finish and most accurate cut is a
   (A) regular-tooth form   (C) claw-tooth form
   (B) hook-tooth form   (D) skip-tooth form

   4.   A   B   C   D

5. For deep cuts in soft material, the best tooth form is
   (A) claw tooth   (C) regular tooth
   (B) buttress tooth   (D) precision tooth

   5.   A   B   C   D

6. The pitch of the blade to be used is determined by the
   (A) r/min of the carrier wheels   (C) best tooth form for the finish required
   (B) thickness of the material to be cut   (D) speed of the feeding process

   6.   A   B   C   D

7. The *minimum* number of teeth in contact with the material being cut should be
   (A) four   (C) two
   (B) three   (D) one

   7.   A   B   C   D

8. The set of a bandsaw blade
   (A) determines the cutting accuracy   (C) is the same as the pitch
   (B) produces the best finish   (D) produces the clearance for the back of
   the blade

   8.   A   B   C   D

9. A raker-set blade is best used for
   (A) general application   (C) workpieces with changing cross sections
   (B) cutting thin sheet metal   (D) cutting light nonferrous metal

   9.   A   B   C   D

10. The best blade set for cutting Bakelite is
    (A) raker   (C) straight
    (B) wave   (D) plain

    10.   A   B   C   D

11. For cutting a .25-in.-thick section, the best pitch to select for speed and efficiency would be
    (A) 8 pitch   (C) 12 pitch
    (B) 10 pitch   (D) 16 pitch

    11.   A   B   C   D

12. The gage of a saw blade is the
    (A) width   (C) width of cut
    (B) thickness   (D) offset of teeth

    12.   A   B   C   D

13. A saw blade having one tooth offset to the right, and one offset to the left, and the third tooth set straight is
    (A) straight set   (C) raker set
    (B) wave set   (D) precision set

    13.   A   B   C   D

continued on next page

14. When selecting a blade for cutting a contour, select
    (A) the narrowest blade
    (B) the widest blade that can cut the smallest radii
    (C) a fine-pitch blade
    (D) a flexible blade

14.   A   B   C   D

Place the name of each illustrated saw-blade type in the proper space in the right-hand column.

15.

16.

17.

18.   **19.**   **20.**

15. _____

16. _____

17. _____

18. _____

19. _____

20. _____

SCORE _____

20

The bandsaw permits cutting flat workpieces close to the desired shape or layout with very little material being wasted in the form of chips. By sawing close to the layout line, further machining operations are performed more quickly.

Place the correct word(s) in the blank space(s) provided at the right-hand side of the page that will make the sentence complete and true.

1. Before mounting a bandsaw blade, always select the correct saw guides for the __?__ of the blade.

        1. _____

2. The __?__ guide should be set about __?__ above the work.

        2. _____

           _____

3. When mounting a bandsaw blade, be sure the teeth point down toward the __?__ .

        3. _____

4. With the gearshift lever in __?__ , turn the upper pulley by __?__ to see that the blade is __?__ properly.

        4. _____

           _____

           _____

5. Set the machine to the proper __?__ for the type of blade and the __?__ to be cut.

        5. _____

           _____

6. Always feed the work with a work-holding jaw or a piece of __?__ .

        6. _____

7. When cutting to a layout line, always cut to within __?__ of the line.

        7. _____

8. When making curved cuts, use as __?__ a blade as possible.

        8. _____

9. The blade length for a two-wheel bandsaw is calculated by adding __?__ the center-to-center distance and the __?__ of one pulley.

        9. _____

           _____

10. Before welding the blade, make sure both ends are __?__ .

        10. _____

11. When welding a blade, adjust the jaw pressure by turning the selector handle for the __?__ of the blade.

        11. _____

12. When annealing the blade, never allow it to get too hot since it may __?__ as it cools.

        12. _____

13. When annealing, heat the weld to a __?__ red color.

        13. _____

14. After annealing, grind the weld to the thickness of the __?__ .

        14. _____

15. After grinding the weld, reanneal the blade to a __?__ color.

        15. _____

16. The best filing speed is between __?__ and __?__ ft/min.

        16. _____

           _____

17. Apply only __?__ work pressure to the file band when filing.

        17. _____

18. Cardboard, cork, and rubber are best cut with a __?__ -edge blade.

        18. _____

19. __?__ -edge blades have a continuous helical cutting edge around the circumference.

        19. _____

SCORE _____

25

TEST 35   Sawing, Filing, and *(Unit 37)*
Additional Band Tools

Decide whether each statement is true or false and circle the letter in the right-hand column that indicates your choice.

1. Always use as narrow a blade as possible.   1.   T   F

2. Drill several holes wherever a sharp turn is required.   2.   T   F

3. A saw blade will wander if the pitch is too coarse.   3.   T   F

4. A blade may not cut due to improper tension.   4.   T   F

5. If too heavy a feed is used, the saw teeth will dull quickly.   5.   T   F

6. The upper saw guide should clear the top of the work by .50 in.   6.   T   F

7. Friction sawing is the fastest method of sawing ferrous metals up to 1-in. thick.   7.   T   F

8. The blade becomes red-hot when friction sawing.   8.   T   F

9. Friction sawing is not suitable for cutting stainless-steel alloys.   9.   T   F

10. Aluminum and brass cannot be cut by friction sawing.   10. T   F

11. High-speed bandsawing is performed at speeds ranging from 2000 to 6000 r/min.   11. T   F

12. Buttress or claw-tooth blades are most efficient for high-speed sawing.   12. T   F

13. The same machine setup is used for high-speed sawing as is used with conventional sawing.   13. T   F

14. Band files are available in only flat and round cross sections.   14. T   F

15. No special attachments are required for band filing.   15. T   F

16. The proper file band to use is indicated on the job selector dial.   16. T   F

17. Knife-edge blades are available only in straight and scalloped edges.   17. T   F

18. Aluminum oxide and silicon carbide are used to make line-grinding bands.   18. T   F

19. Diamond-edge blades have diamond particles fused to the edges of the saw teeth.   19. T   F

20. Electro-band machining requires a coolant.   20. T   F

SCORE _____

20

Metal-Cutting Saws Review Test *(Section 9)*

PART 1

Place the correct word(s) in the blank space(s) provided at the right-hand side of the page that will make the sentence complete and true.

1. A __?__ -pitch blade is recommended for general-purpose work on cutoff saws.   1. _____

2. Bandsaw blades have only the __?__ hardened.   2. _____

3. Too __?__ a blade speed or too __?__ a feeding pressure will dull the saw teeth quickly.   3. _____
_____

4. Before setting work in the vise of a cutoff saw, be sure that the solid jaw is __?__ .   4. _____

5. When holding short work in a vise of a cutoff saw, always place a short piece of the __?__ thickness in the __?__ end of the vise.   5. _____
_____

6. Before taking a cut using a cutoff saw, always adjust the roller guide __?__ so that they just __?__ the work.   6. _____
_____

7. Before taking a cut using a cutoff saw, always __?__ the length of the material and see that the vise is __?__ .   7. _____
_____

8. When sawing material, always be sure that at least __?__ teeth are touching the work.   8. _____

9. Before mounting a bandsaw blade, always select the correct saw guides for the __?__ of the blade.   9. _____

10. When mounting a bandsaw blade, always point the teeth toward the __?__ .   10. _____

11. With the gearshift lever in __?__ , turn the upper pulley by __?__ to see that the blade is __?__ properly.   11. _____
_____
_____

12. When cutting to a layout line, always cut to within __?__ in. of the line.   12. _____

13. The blade length for a two-wheel bandsaw is calculated by adding __?__ the center-to-center distance and the __?__ of the pulley.   13. _____
_____

14. Before welding a blade, make sure both ends are __?__ .   14. _____

15. After annealing, grind the weld to the thickness of the __?__ .   15. _____

16. After grinding the weld, re-anneal the blade to a __?__ color.   16. _____

17. Never attempt to cut a small radius with too __?__ a blade.   17. _____

18. Apply only __?__ work pressure to the file band when filing.   18. _____

19. Cardboard, cork, and rubber are best cut with a __?__ -edge blade.   19. _____

PART 2

Select the most appropriate answer for each question and circle the letter in the right-hand column that indicates your choice.

20. A bandsaw blade, when properly selected, will hold its sharpness because   20.   A   B   C   D
    (A) of the hardness of the saw teeth
    (B) wear is distributed over many teeth
    (C) less horsepower is required to drive the blade
    (D) of the tooth set

21. Which of the following statements is *not true* regarding contour bandsaws?   21.   A   B   C   D
    (A) Machining geometry is not restricted.
    (B) Only simple fixturing is required.
    (C) Cutting force holds the work on the table.
    (D) Material removed is reduced to chips.

22. The saw blade that produces the best finish and most accurate cut is a   22.   A   B   C   D
    (A) regular-tooth form
    (B) hook-tooth form
    (C) claw-tooth form
    (D) skip-tooth form

continued on next page

23. For deep cuts in soft material, the best tooth form is a   23.   A   B   C   D
(A) claw tooth   (C) precision tooth
(B) buttress tooth   (D) regular tooth

24. The pitch of a blade is determined by the   24.   A   B   C   D
(A) r/min of the carrier wheels   (C) gage of the saw
(B) thickness of the material being cut   (D) rate of feed

25. The set of the saw blade   25.   A   B   C   D
(A) determines the accuracy of the cut   (C) is the same as the pitch
(B) produces a finer finish   (D) produces clearance for back of blade

26. A raker-set blade is used for   26.   A   B   C   D
(A) general applications   (C) cutting thin sheet metal
(B) work pieces with changing cross   (D) cutting light nonferrous metal
     sections

27. The gage of a saw blade is the   27.   A   B   C   D
(A) width   (C) width of cut
(B) thickness   (D) offset of teeth

28. A saw blade having one tooth offset to the right, and one offset to the left, and the third   28.   A   B   C   D
set straight is
(A) straight set   (C) raker set
(B) wave set   (D) precision set

PART 3

Decide whether each statement is true or false and circle the letter in the right-hand column that indicates your choice.

29. Always use as wide a blade as possible for any sawing job.   29.   T   F

30. The precision tooth has a 30° rake angle.   30.   T   F

31. The buttress tooth is faster cutting than the claw tooth.   31.   T   F

32. The buttress-tooth form is different from the regular-tooth form.   32.   T   F

33. A raker-set blade has one tooth offset to the right and the next to the left.   33.   T   F

34. The gage of a saw blade is the thickness of the saw blade measured over the offset of the teeth.   34.   T   F

35. Notching is an economical method of metal removal.   35.   T   F

36. A saw blade will wander due to improper blade tension.   36.   T   F

37. Knife-edge blades are available only with straight, wavy, and scalloped edges.   37.   T   F

38. The spring-tempered spiral-edge blade should not be used to cut metals.   38.   T   F

PART 4
Place the name of each illustrated saw-blade type or sawing operation in the proper space in the right-hand column.

39. _____

40. _____

41. _____

42. _____

43. _____

SCORE _____

50

39.

40.

41.   42.   43.

TEST 36   Drill Presses and *(Units 38, 39)*
Accessories

The drill press is one of the most common machine tools used in metal-working shops for producing, forming, and finishing holes. Although there are many types of drill presses, they all operate on the principle of forcing a revolving cutting tool into a workpiece.

Place the correct word(s) in the blank space(s) provided at the right-hand side of the page that will make the sentence complete and true.

1. __?__ is the operation of producing a hole by removing metal from a solid mass with a cutting tool called a __?__ .

2. Tapping is the operation of cutting internal __?__ in a hole.

3. __?__ is the operation of __?__ and producing a smooth round hole from a previously drilled or bored hole.

4. The size of a drill press may be designated as the distance from the center of the __?__ to the __?__ of the machine.

5. A __?__ drill press is one that has a hand-feed mechanism and enables the operator to __?__ how the drill is cutting.

6. __?__ drills are drilling machines with more than one __?__ or drilling head mounted on a single table.

7. Name the six drill press parts indicated by the letters.

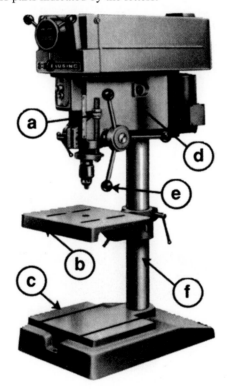

1. drilling
   twist drill
2. threads
3. Reaming
   Sizing
4. spindle
   column
5. Sensitive
   feel
6. multiple spindle
   spindle

7.(a) depth stop
  (b) table
  (c) base
  (d) drilling head
  (e) hand feed lever
  (f) column

continued on next page

8. A drill ___?___ can be used to adapt a cutting tool shank to fit a larger machine spindle taper.

9. Three types of vises used on drill presses are ___?___ , ___?___ , and ___?___ .

10. ___?-?___ are used to hold round work, ___?___ ___?___ are used to provide support for strap clamps, and ___?___ ___?___ are used to produce drilling of identical parts.

11. When using strap clamps to fasten work to the table, the step blocks should be slightly ___?___ than the workpiece.

8. _Sleeve_

9. _drill_
contour
Angle

10. _v-blocks_
step-blocks
blocks
drill
jigs

11. _higher_

SCORE _____
                    25

TEST 37   Twist Drills *(Unit 40)*

The twist drill, one of the most common and probably the most abused of the metal-cutting tools, is used to produce holes in metal or other materials. Care, proper selection, and an understanding of its limitations can prolong the life and accuracy of the twist drill.

Select the most appropriate answer for each question and circle the letter in the right-hand column that indicates your choice.

1. The speed at which carbide drills can be operated, compared with high-speed drills, is
   - (A) half speed
   - (B) double speed
   - (C) triple speed
   - (D) one-quarter speed

   1.   A   B   C   D

2. Which of the following statements does *not* apply to cemented-carbide drills?
   - (A) They operate at high speeds.
   - (B) Cutting edges do not wear rapidly.
   - (C) They can withstand higher heat.
   - (D) They operate at the same speed as high-speed drills.

   2.   A   B   C   D

3. Generally, drills with straight shanks are manufactured up to a diameter of
   - (A) 1/4 in.
   - (B) 1/2 in.
   - (C) 5/8 in.
   - (D) 7/16 in.

   3.   A   B   C   D

4. Which of the following statements is *not true* with respect to the flutes of a drill?
   - (A) They form the cutting edge.
   - (B) They admit cutting fluid.
   - (C) They are manufactured only in helical form.
   - (D) They allow chips to escape.

   4.   A   B   C   D

5. The margin of a drill is
   - (A) a narrow raised section on the body next to the flute
   - (B) the clearance behind the cutting edge
   - (C) the distance across the top of the flutes
   - (D) another name for the flutes

   5.   A   B   C   D

6. Which of the following statements is *not true* regarding the web of a drill?
   - (A) It is the same thickness throughout the length of the drill.
   - (B) It runs the full length of the flutes.
   - (C) It is the center portion of the drill separating the flutes.
   - (D) It forms a chisel edge at the point of the drill.

   6.   A   B   C   D

7. The average lip clearance ground on a drill should be
   - (A) 5 to 7°
   - (B) 8 to 12°
   - (C) 3 to 5°
   - (D) 2 to 3°

   7.   A   B   C   D

8. Fractional drill sizes range in diameter from 1/64 to 3 1/4 in. by increments of
   - (A) .010 in.
   - (B) 1/32 in.
   - (C) 1/64 in.
   - (D) .050 in.

   8.   A   B   C   D

9. A #27 drill, as compared with a #26 drill, is
   - (A) larger in diameter
   - (B) greater in diameter by .010 in.
   - (C) smaller in diameter
   - (D) smaller in diameter by .010 in.

   9.   A   B   C   D

10. Which of the following statements is *not true* with respect to the letter-size drills?
    - (A) They range from A to Z.
    - (B) A is the smallest; Z is the largest.
    - (C) A is the largest; Z is the smallest.
    - (D) Letter E is 1/4-in. diameter.

    10.   A   B   C   D

11. High-helix drills are designed to
    - (A) drill deep holes in aluminum
    - (B) enlarge punched holes
    - (C) have two oil holes running the length of the drill
    - (D) drill holes up to several feet deep

    11.   A   B   C   D

12. Oil hole drills may
    - (A) have one or two holes running the length of the drill
    - (B) use compressed air as a coolant
    - (C) use oil as a coolant
    - (D) all of these

    12.   A   B   C   D

continued on next page

13. A general-purpose drill for drilling most metals has an included-point angle of   13.   A   B   C   D
    (A) 150°   (C) 118°
    (B) 130°   (D) 90°

14. On the pedestal grinder, how far should the tool rest be set from the wheel?   14.   A   B   C   D
    (A) .18 in.   (C) .12 in.
    (B) .25 in.   (D) .06 in.

15. When the lip of the drill is against the face of the grinding wheel, the shank should be   15.   A   B   C   D
    (A) lowered and not twisted   (C) held horizontal and twisted
    (B) lowered and twisted   (D) raised and twisted

16. In regard to grinding a drill, which of the following statements is *not true*?   16.   A   B   C   D
    (A) Hold the shank slightly lower than   (C) Hold the cutting edge of the drill
        the point.   parallel to the tool rest.
    (B) Hold the drill at 118° to the face of   (D) Grind a lip clearance of about 10°
        the wheel.

17. Which of the following statements is *not true*? Too much lip clearance   17.   A   B   C   D
    (A) weakens the cutting edge   (C) damages the cutting edges
    (B) requires more drilling pressure   (D) causes the lip to chip

18. A drill that is ground with lips of unequal length will drill a hole that is   18.   A   B   C   D
    (A) oversized   (C) bell-mouthed
    (B) tapered   (D) all of these

19. A drill that has been ground with unequal lip angles will drill a hole that is   19.   A   B   C   D
    (A) oversized   (C) bell-mouthed
    (B) tapered   (D) not concentric

20. Which of the following statements is *not true* in regard to incorrect drill-point angles and   20.   A   B   C   D
    clearances?
    (A) They may produce oversize holes.   (C) They may alter the shape of the chip.
    (B) They may produce undersize holes.   (D) They will affect the pressure required
                                       to feed the drill.

SCORE _____
                                              20

TEST 38   Drilling Speeds and Feeds *(Unit 41)*

The speed and feed at which a drill is operated affect the life of the drill. Constant attention to both of these factors is required by the drill-press operator.

Decide whether each statement is true or false and circle the letter in the right-hand column that indicates your choice.

1.	Too slow a cutting speed will dull the drill quickly.	1.	T	F
2.	Cutting speed refers to peripheral speed.	2.	T	F
3.	Drilling speed is affected by the efficiency of the cutting fluid.	3.	T	F
4.	The cutting speed of a large drill is greater than that of a smaller drill.	4.	T	F
5.	The cutting speed of any diameter high-speed drill is the same when machine steel is being cut.	5.	T	F
6.	The speed at which a drill revolves (r/min) is determined only by the material being cut.	6.	T	F
7.	A drill that is run faster than the recommended cutting speed will cut more efficiently.	7.	T	F
8.	*Feed* may be defined as the distance the drill advances into the work in 1 min.	8.	T	F
9.	Too light a feed will cause the cutting edges of the drill to dull quickly.	9.	T	F
10.	Small drills require a higher feed per revolution than larger ones.	10.	T	F

SCORE _____

10

TEST 39   Drilling Holes (*Unit 42*)

The practice of drilling holes was probably one of the first cutting-tool operations ever performed by humans. Although this practice is relatively simple, great care must be exercised in the setup and the drilling of the work, with a special emphasis on safety.

Select the most appropriate answer for each question and circle the letter in the right-hand column that indicates your choice.

1. A hole location is generally spotted using a
   (A) number drill
   (B) letter drill
   (C) center drill
   (D) 1/4-in. drill

   1.   A   B   C   D

2. When spotting a hole location,
   (A) hold the work in a vise that is clamped to the table
   (B) clamp the work to the table
   (C) clamp the work on the table or parallels
   (D) do not clamp the work or vise

   2.   A   B   C   D

3. When drilling work that is not clamped to the table, prevent it from spinning by
   (A) the column of the drill press
   (B) a clamp or table stop on the left-hand side of the table
   (C) a clamp or table stop on the right-hand side of the table
   (D) holding it by hand

   3.   A   B   C   D

4. If possible, flat work held in a vise for drilling should be mounted on
   (A) the vise base
   (B) parallels
   (C) a wooden block
   (D) top of the vise jaws

   4.   A   B   C   D

5. Accurate hole locations should be marked with a
   (A) prick punch
   (B) scribed and punched circle to show the exact location of the hole
   (C) center punch
   (D) scribed and punched test circle and a proof circle

   5.   A   B   C   D

6. If a hole is to be drilled to an accurate location, it should be
   (A) center-drilled
   (B) drilled to one-half the depth of the drill point
   (C) moved over, if necessary, using a cape chisel
   (D) all of these

   6.   A   B   C   D

7. Which of the following statements is *not true* regarding pilot holes?
   (A) They provide a channel for coolant.
   (B) They should be slightly larger than the size of the web.
   (C) They relieve some of the drilling pressure when using large drills.
   (D) They prevent the drill from wandering.

   7.   A   B   C   D

8. When the drill point is about to break through the workpiece, the pressure on the drill should
   (A) remain the same
   (B) be increased slightly
   (C) be decreased
   (D) be increased rapidly

   8.   A   B   C   D

9. If a pilot hole is too large, it may
   (A) reduce chattering
   (B) cause the hole to be drilled out-of-round
   (C) damage the bottom of the hole
   (D) all of these

   9.   A   B   C   D

10. When a center drill is used in spotting a hole, the drill should revolve at about
    (A) 400 r/min and be fed in to one-half the size of the hole to be drilled
    (B) 800 r/min and be fed in for the full depth of the tapered section
    (C) 1000 r/min and be fed in to the full size of the hole to be drilled
    (D) 1500 r/min and be fed in for about one-third the length of the tapered section

    10.   A   B   C   D

SCORE _____

10

TEST 40   Reamers *(Unit 43)*

A *reamer* is a rotary cutting tool having two or more flutes that is used to finish previously drilled or bored holes to accurate dimensions.

Place the correct word(s) in the blank space(s) provided at the right-hand side of the page that will make the sentence complete and true.

1. The main parts of a reamer are the __?__ , the __?__ , and the __?__ .

    1. _____

    _____

    _____

2. The margin is ground with body __?__ to reduce __?__ when the reamer is cutting.

    2. _____

    _____

3. The two types of reamers are __?__ and __?__ .

    3. _____

    _____

4. __?__ fluted reamers are suited for finishing holes having slots or keyways.

    4. _____

5. Rose reamers cut on the __?__ only.

    5. _____

6. Fluted chucking reamers have __?__ teeth than rose reamers.

    6. _____

7. Adjustable reamers are adjustable to about __?__ in. over or under the reamer size.

    7. _____

8. Hand reamers may be identified by a __?__ on the end of the shank.

    8. _____

9. Never turn a reamer __?__ .

    9. _____

10. The allowance for machine reaming a .250-in.-diameter hole should be no more than __?__ in.

    10. _____

11. The allowance for machine reaming a .750-in.-diameter hole should be no more than __?__ in.

    11. _____

12. The reaming speed is generally about one- __?__ of that used for drilling.

    12. _____

13. The shank end of a hand reamer is supported by a __?__ center when hand reaming is done on a drill press.

    13. _____

14. When hand reaming, always turn the reamer in a __?__ direction.

    14. _____

15. When removing a reamer, always turn it in a __?__ direction.

    15. _____

16. Always store reamers in __?__ compartments or containers.

    16. _____

SCORE _____

20

TEST 41   Drill-Press Operations *(Unit 44)*

The drill press, like other machine tools, is capable of performing several operations other than drilling and reaming. Although many of these operations may be done on other machines or by hand, it is often more convenient and more accurate to do them on the drill press.

Place the correct word(s) in the blank space(s) provided at the right-hand side of the page that will make the sentence complete and true.

1. Ductile materials can be form threaded using a __?__ __?__ .

    1. _____

    _____

2. After the hole has been drilled, to ensure tap alignment, the work should not be __?__ .

    2. _____

3. When tapping by hand in the drill press, use a __?__ center to keep the tap aligned.

    3. _____

4. When cutting a right-hand thread, turn the tap in a __?__ direction.

    4. _____

5. When a tapping attachment is being used, it is better to use a __?__ -fluted or __?__ -fluted machine or __?__ tap.

    5. _____

    _____

    _____

6. Machine taps clear the __?__ better than hand taps.

    6. _____

7. When a tapping attachment is being used, the rotation of the tap may be stopped by __?__ the downward pressure on the spindle feed handle.

    7. _____

8. *Counterboring* is the operation of enlarging the __?__ of a hole to a given __?__ .

    8. _____

    _____

9. The counterbore is guided into the hole by a __?__ on the end of the cutter.

    9. _____

10. Counterboring is performed at about one- __?__ drilling speed.

    10. _____

11. Countersinking produces a __?__ -shaped hole at the top of the drilled hole.

    11. _____

12. The diameter of a countersink hole may be checked by placing the head of an inverted __?__ -head machine screw in the top of the hole.

    12. _____

13. Most flat-head machine screws require an __?__ ° countersink.

    13. _____

14. The speed for countersinking is about one- __?__ drilling speed.

    14. _____

15. If several holes must be countersunk to the same diameter, the __?__ stop should be set on the drill press.

    15. _____

16. Three common methods of transferring hole locations are spotting with a twist drill, __?__ , and __?__ .

    16. _____

    _____

17. Drill jigs eliminate the need to __?__ hole locations.

    17. _____

18. A drill jig uses __?__ bushings to guide and position the drill.

    18. _____

SCORE _____

25

Drilling Machines and Drilling *(Section 10)*
Review Test

PART 1

Place the name of each illustrated drill press, accessory, or operation in the proper space in the right-hand column.

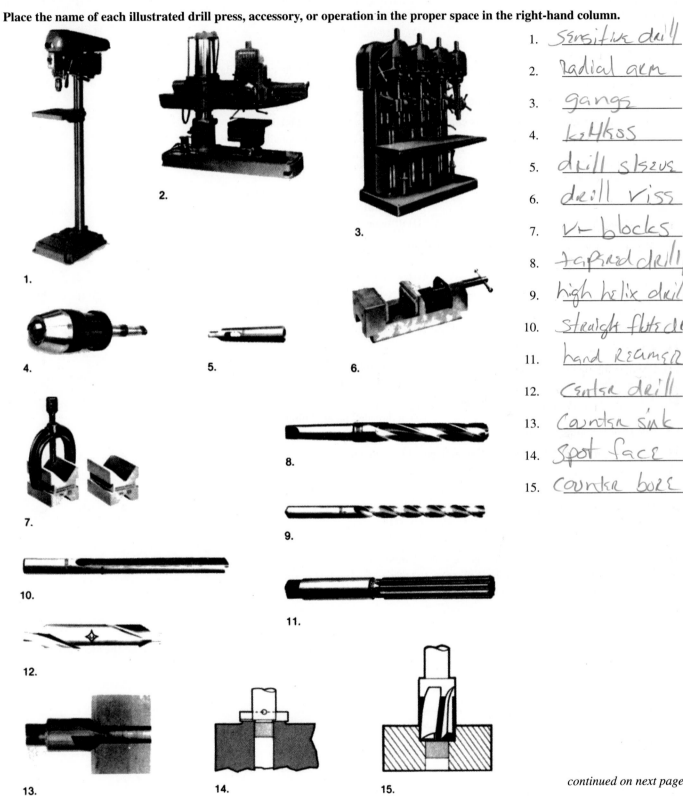

1. Sensitive drill
2. Radial arm
3. gangs
4. k.4k55
5. drill sleeve
6. drill viss
7. v- blocks
8. tapired drillbt
9. high helix drill
10. straight flute drill
11. hand REAMER
12. center drill
13. counter sink
14. spot face
15. counter bore

continued on next page

PART 2

Select the most appropriate answer for each question and circle the letter in the right-hand column that indicates your choice.

16. The speed at which carbide drills can be operated, compared with high-speed drills, is   16.   A   B   C   D
    (A) half speed   (C) triple speed
    (B) double speed   (D) one-quarter speed

17. Generally, drills with straight shanks are manufactured up to a diameter of   17.   A   B   C   D
    (A) 1/4 in.   (C) 5/8 in.
    (B) 1/2 in.   (D) 7/16 in.

18. The margin of a drill is   18.   A   B   C   D
    (A) a narrow, raised section on the body   (C) the clearance behind the cutting edge
    next to the flute   (D) another name for flutes
    (B) the distance across the top of the flutes

19. Which of the following statements is *not true* with regard to the flutes of a drill?   19.   A   B   C   D
    (A) They allow chips to escape.   (C) They are made only in helical form.
    (B) They admit cutting fluid.   (D) They form the cutting edges.

20. Which of the following statements is *not true* regarding the web of a drill?   20.   A   B   C   D
    (A) It is the same thickness for the length   (C) It forms the chisel edge at the point of
    of the drill.   the drill.
    (B) It should be thinned occasionally.   (D) It separates the flutes.

21. The average lip clearance ground on a general-purpose drill should be   21.   A   B   C   D
    (A) 5 to 7°   (C) 3 to 5°
    (B) 8 to 12°   (D) 2 to 3°

22. Fractional drill sizes range in diameter from 1/64 to 3 1/4 in. and occur in increments of   22.   A   B   C   D
    (A) .010 in.   (C) 1/64 in.
    (B) 1/32 in.   (D) .050 in.

23. A hand reamer may be identified by   23.   A   B   C   D
    (A) spiral flutes   (C) straight flutes
    (B) a square on the shank   (D) a tapered shank

24. An adjustable reamer is enlarged by means of   24.   A   B   C   D
    (A) two adjusting nuts   (C) a threaded tapered plug
    (B) a square on the shank   (D) a tapered pin

25. The teeth on the end of a hand reamer are tapered for a length of   25.   A   B   C   D
    (A) one-third the length of the reamer   (C) .50 in.
    (B) 1 in.   (D) the diameter of the reamer

26. To hand-ream a 3/4-in. hole, first drill or bore the hole to   26.   A   B   C   D
    (A) .750   (C) .745
    (B) .700   (D) .735

27. Which of the following statements is *not true* regarding machine taps?   27.   A   B   C   D
    (A) They usually have two or three flutes.   (C) They are supplied in sets of three.
    (B) They are used with a tapping attachment.   (D) They clear chips quickly.

PART 3

Place the correct word(s) in the blank space(s) provided at the right-hand side of the page that will make the sentence complete and true.

28. The sensitive drill press does not have an automatic __?__ .   28. _____

29. A radial drill press is designed to machine __?__ work.   29. _____

30. The size of a drilling machine may be determined by the distance from the __?__ to the   30. _____
    center of the __?__ .
                                                                         _____

31. Providing a recess at the top of a hole to accommodate a bolt head is called __?__ .   31. _____

32. A #1 drill is _____ than a #40 drill.   32. _____

33. A rose chucking reamer has __?__ teeth than a fluted chucking reamer.   33. _____

34. A reamer having __?__ flutes should be used when reaming a hole that has a keyway.   34. _____

35. Never set speeds or adjust the work unless the drill is __?__ .   35. _____

36. A drill should be reground at the first sign of __?__ .   36. _____

37. Before inserting a tapered drill shank, always __?__ the shank and the spindle hole.   37. _____

38. A __?__ chisel is used to draw a drill over to the correct location.   38. _____

39. A __?__ should be fastened to the left side of the table to prevent the work from spinning. 39. _____

PART 4

Decide whether each statement is true or false and circle the letter in the right-hand column that indicates your choice.

40. Carbide drills are operated at one-half the speed of high-speed drills.   40. T   F

41. The web of the drill tapers for the length of the body of a drill.   41. T   F

42. A core drill has coolant holes running from the shank to the cutting edges of the drill.   42. T   F

43. The cutting speed of a large drill is greater than that of a small drill.   43. T   F

44. Machine reamers are supplied only with tapered shanks.   44. T   F

45. An expansion reamer is adjusted oversize by a threaded taper plug.   45. T   F

46. A pilot hole is required before drilling any hole to size.   46. T   F

47. The rate of feed for drilling is calculated using the formula $\dfrac{4 \times CS}{D}$.   47. T   F

48. A center drill has an included angle of 60°.   48. T   F

49. When buffing a workpiece, never hold it with a cloth.   49. T   F

SCORE _____

50

TEST 42   Engine Lathe Sizes, Parts, and Safety   *(Units 45, 48)*

The lathe is one of the most versatile machines in the machine shop. It operates on the same principle as the early hand- or foot-operated potter's wheel, which revolved lumps of clay to shape it into pottery. Turning, tapering, knurling, threading, drilling, boring, and reaming are some of the most common operations performed on a lathe. With the addition of computer numerical control, the lathe has been developed into a highly sophisticated machine, capable of performing many operations with much greater speed and accuracy than machines of even a decade ago.

Select the most appropriate answer for each question and circle the letter in the right-hand column that indicates your choice.

1. Lathe sizes are usually designated by the   1.   A   B   C   D

   (A) length of the bed   (C) length of work that can be held
   (B) diameter of work that can be swung   between centers
                                            (D) diameter and length of work that can
                                                be accommodated

2. The carriage may be advanced automatically for turning operations by the   2.   A   B   C   D

   (A) automatic feed lever   (C) lead screw
   (B) split-nut lever   (D) feed-directional lever

3. The feed-reverse lever reverses the movement of the   3.   A   B   C   D

   (A) longitudinal feed   (C) angular feed
   (B) crossfeed   (D) both A and B

4. The quick-change gearbox provides the drive for the   4.   A   B   C   D

   (A) lead screw   (C) lead screw and feed rod
   (B) feed rod   (D) headstock spindle

5. Which of the following lathe items is *not* fastened to the carriage?   5.   A   B   C   D

   (A) saddle   (C) cross slide
   (B) apron   (D) quick-change gearbox

6. The apron handwheel usually moves the carriage by means of   6.   A   B   C   D

   (A) spur gears   (C) rack and (pinion) gear
   (B) bevel gears   (D) miter gears

7. When thread cutting, be sure the feed-change lever is   7.   A   B   C   D

   (A) engaged   (C) set for longitudinal feed
   (B) in neutral position   (D) set for crossfeed

8. The feed rod drives the carriage by means of   8.   A   B   C   D

   (A) bevel gears   (C) a clutch
   (B) a split nut   (D) an acme thread

9. Which of the following parts is *not* found on the tailstock?   9.   A   B   C   D

   (A) live center   (C) spindle-binding lever
   (B) dead center   (D) handwheel

10. The compound rest may be rotated   10. A   B   C   D

    (A) 180°   (C) 360°
    (B) 60°   (D) 90°

Place the correct word(s) in the blank space(s) provided at the right-hand side of the page that will make the sentence complete and true.

11. Always wear approved __?__ glasses when operating a lathe.   11. _____

12. Never attempt to __?__ any machine until you are __?__ with its operation.   12. _____

_____

13. Never wear loose __?__ , watches, or __?__ when operating a lathe.

13. _____

_____

14. Always stop the lathe before __?__ the work.

14. _____

15. Use a __?__ to remove chips.

15. _____

16. __?__ and __?__ left on the floor can cause falls.

16. _____

_____

17. Remove sharp __?__ from the machine when polishing, filling, cleaning, or making adjustments.

17. _____

SCORE _____

20

TEST 43   Lathe Accessories (*Unit 46*)

Lathe accessories increase the versatility of the machine. They include various devices used to hold, support, and drive the work, as well as those used for holding and mounting various cutting tools.

Select the most appropriate answer for each question and circle the letter in the right-hand column that indicates your choice.

1. A revolving tailstock center
   (A) must be oiled regularly
   (B) is affected by work expansion
   (C) does not require lubrication
   (D) is used only in the headstock

   1.   A   B   C   D

2. A self-driving center is
   (A) ordinarily used with soft material
   (B) used in the tailstock
   (C) used for multiple-turning operations
   (D) grooved on the insertion taper

   2.   A   B   C   D

3. The three-jaw universal chuck can be used to hold
   (A) square work
   (B) octagonal and round work
   (C) octagonal work
   (D) hexagonal and round work

   3.   A   B   C   D

4. Which of the following statements does *not* apply to a universal chuck?
   (A) Its jaws are actuated by a scroll plate.
   (B) It holds work extremely accurately.
   (C) It is made in various sizes from 4 to 16 in.
   (D) It is provided with two sets of jaws.

   4.   A   B   C   D

5. The four-jaw independent chuck can hold
   (A) square work only
   (B) round work only
   (C) irregular-shaped work only
   (D) all of these

   5.   A   B   C   D

6. Which of the following does *not* apply to a headstock spindle?
   (A) threaded
   (B) tapered
   (C) splined
   (D) cam lock

   6.   A   B   C   D

7. The chuck that is best suited for holding round, square, and irregular-shaped pieces is the
   (A) universal chuck
   (B) independent chuck
   (C) magnetic chuck
   (D) collet chuck

   7.   A   B   C   D

8. Which of the following diameters should *not* be used with a 1/2-in. collet chuck?
   (A) .498 in.
   (B) .501 in.
   (C) .499 in.
   (D) .490 in.

   8.   A   B   C   D

9. When work is of such nature that it must be mounted off center on a faceplate,
   (A) lathe speed should be reduced to 50 r/min
   (B) counterbalances should be used
   (C) some alternate method should be selected
   (D) it is strictly a boring-mill operation

   9.   A   B   C   D

10. A steadyrest is
    (A) clamped to the carriage
    (B) restricted to long work held in a chuck
    (C) aligned by the ways of the lathe
    (D) restricted to long work held between centers.

    10.   A   B   C   D

Decide whether each statement is true or false and circle the letter in the right-hand column that indicates your choice.

11.	A straight-tail lathe dog is best suited for precision turning between centers.	11.	T F
12.	A standard-type toolholder has a back rake of 15 to 20°.	12.	T F
13.	A left-hand offset toolholder is designed to machine work from right to left.	13.	T F
14.	A right-hand offset toolholder is designed to machine work close to the chuck.	14.	T F
15.	Carbide cutting tools should have a back-rake angle of 15 to 20°.	15.	T F
16.	Form-relieved threading tools must only be ground on the top of the cutting surface.	16.	T F
17.	The standard, or round-type, tool post is best suited for carbide cutting tools.	17.	T F
18.	Most turret-type tool posts can accommodate only two different cutting tools.	18.	T F
19.	Quick-change toolposts usually have a dovetail for accurate repositioning of the cutting tool.	19.	T F
20.	Accuracy of size is maintained with a quick-change toolpost by setting the cutting tool to the work and checking the size after each cut.	20.	T F

SCORE _____

20

TEST 44 Cutting Speeds and Feeds (Unit 47)

The use of the proper cutting speed is very important since too slow a cutting speed will waste valuable time. Too fast a cutting speed will cause the cutting tool to dull quickly and time will be wasted in regrinding and resetting the tool.

Select the most appropriate answer for each question and circle the letter in the right-hand column that indicates your choice.

1. In order to machine the work in the shortest time possible, you should
    - (A) increase the spindle speed and decrease the feed
    - (B) decrease the cutting speed and the feed
    - (C) use the recommended cutting speed and feed
    - (D) decrease the spindle speed to prolong the life of the cutting tool

    1. A   B   C   D

2. If the cutting speed is increased for the workpiece being machined, the
    - (A) output will be greater
    - (B) cutting-tool wear will be compensated by increased production
    - (C) cutting-tool edge will break down, resulting in loss of time
    - (D) cutting tool will have to be changed more

    2. A   B   C   D

3. The *cutting speed* may be defined as
    - (A) the rate at which a point on the work radius passes the cutting tool in 1 min
    - (B) the spindle speed of the lathe
    - (C) the rate at which a point on the work circumference passes the cutting tool in 1 min
    - (D) feet or meters per min

    3. A   B   C   D

4. The r/min at which to set the lathe is calculated by which of the following formulas?
    - (A) $\dfrac{CS}{4 \times D}$
    - (B) $\dfrac{D}{4 \times CS}$
    - (C) $\dfrac{4 \times CS}{\pi D}$
    - (D) $\dfrac{4 \times CS}{D}$

    4. A   B   C   D

5. The r/min for rough turning a 3-in.-diameter machine-steel shaft (CS 90) using a high-speed steel cutting tool would be
    - (A) 120
    - (B) 100
    - (C) 95
    - (D) 80

    5. A   B   C   D

6. The r/min for finish turning a 2-in.-diameter bronze shaft (CS 100) using a high-speed steel cutting tool would be
    - (A) 240
    - (B) 200
    - (C) 300
    - (D) 120

    6. A   B   C   D

7. The r/min for finish turning an 80-mm-diameter aluminum shaft (CS 93) using a high-speed steel cutting tool would be
    - (A) 186
    - (B) 279
    - (C) 372
    - (D) 465

    7. A   B   C   D

8. *Feed* may be defined as the
    - (A) distance the cutting tool advances along the work for each revolution
    - (B) length of the cutting removed in 1 min
    - (C) distance the cutting tool travels in 1 min
    - (D) length of cutting removed in one revolution

    8. A   B   C   D

9. If a .015-in. feed is set on the lathe, the number of spindle revolutions required to move the cutting tool 2.250 in. along the work would be
    - (A) 100
    - (B) 125
    - (C) 137
    - (D) 150

    9. A   B   C   D

10. The machining time is calculated by which of the following formulas?
    - (A) $\dfrac{\text{length of cut} \times \text{feed}}{\text{r/min}}$
    - (B) $\dfrac{\text{length of cut}}{\text{feed} \times \text{r/min}}$
    - (C) $\dfrac{\text{length of cut}}{\text{r/min}}$
    - (D) $\dfrac{\text{length of cut}}{\text{feed}}$

    10. A   B   C   D

11. The machining time required to take a finish cut off a 10-in.-long piece of 2-in.-diameter   11.   A   B   C   D
machine steel (CS 100) using a feed of .010 in. would be
  (A) 5 min   (C) 7 min
  (B) 6 min   (D) 8 min

From the chart, select the position of the levers for the following and place the proper letters or number in the space provided in the right-hand column.

LEVERS	ENGLISH - THREADS PER INCH					METRIC-PITCH IN M/M			
	SLIDING FEEDS IN THOUSANDTHS					SURFACING ¼		SLIDING	
D **B**	60	56	52	48	44	40	38	36	32
	·5 M/M	·005	·005	·006	·006	·75 M/M	·007	·008	·009
C **B**	30	28	26	24	22	20	19	18	16
	1 M/M	·010	·011	1·25 M/M	·013	1·5 M/M	·015	·016	·017
D **A**	15	14	13	12	11	10	9½	9	8
	2 M/M	·020	·021	2·5 M/M	·025	3 M/M	·029	·031	·034
C **A**	·7½	7	6½	6	5½	5	4¾	4½	4
	4 M/M	·039	·042	5 M/M	·050	6 M/M	·058	·061	·068

12. .010 feed

13. .007 feed

14. 2.5 mm feed

12. lever 1 position   _____

lever 2 position   _____

lever 3 row   _____

13. lever 1 position   _____

lever 2 position   _____

lever 3 row   _____

14. lever 1 position   _____

lever 2 position   _____

lever 3 row   _____

SCORE   _____

20

TEST 45   Preparing the Lathe for *(Unit 49)* Machining Between Centers

Because a large proportion of work in school shops is machined between centers, the proper care and alignment of centers is very important to the student. Of equal importance is the proper setup of the workpiece and the cutting tool. All of these factors will affect the quality of the work produced.

Select the most appropriate answer for each question and circle the letter in the right-hand column that indicates your choice.

1. Headstock center runout may be caused by   1. A   B   C   D
   (A) dirt or cuttings in the spindle   (C) burrs in the spindle
   (B) burrs on the center sleeve   (D) all of these

2. The taper in the headstock spindle should be cleaned by   2. A   B   C   D
   (A) holding a cloth in the spindle while the   (C) pushing a cloth into the hole with a
       spindle is revolving slowly   stick or rod
   (B) running your finger around the taper   (D) using an air hose

3. When mounting a lathe center in a spindle,   3. A   B   C   D
   (A) push it in slowly by hand   (C) insert it with a quick snap
   (B) push it into the taper and tap it with a   (D) oil the center first
       hammer

4. When removing a live center using a knockout bar, you should   4. A   B   C   D
   (A) let it fall onto a wooden tray   (C) hold it by hand
   (B) let it fall into the chip tray   (D) wrap a cloth around it and hold it by
                                                 hand

5. If the lathe centers are not aligned, the workpiece will be   5. A   B   C   D
   (A) tapered   (C) undersize
   (B) stepped   (D) out of round

6. The fastest and most accurate way to align the lathe centers is by   6. A   B   C   D
   (A) the trial-cut method   (C) using a dial indicator and test bar
   (B) using the tailstock graduations   (D) eye

7. When aligning the tailstock, first loosen the   7. A   B   C   D
   (A) tailstock spindle clamp   (C) front adjusting screw
   (B) tailstock clamp nut or lever   (D) back adjusting screw

8. After a trial cut has been taken, the readings on the workpiece are 1.250 at the tailstock   8. A   B   C   D
   end and 1.240 at the headstock end. To align the centers, you should adjust the tailstock
   (A) .005 toward the cutting tool   (C) .010 toward the cutting tool
   (B) .005 away from the cutting tool   (D) .010 away from the cutting tool

9. When aligning the centers using a dial indicator, always set the plunger   9. A   B   C   D
   (A) in a horizontal position   (C) parallel to the ways of the lathe
   (B) on center   (D) all of these

10. When centers are being aligned using a test bar and an indicator, the indicator reading is 0   10. A   B   C   D
    at the tailstock end and + 20 at the headstock end. To align the centers, you should adjust
    the tailstock
    (A) .010 toward the cutting tool   (C) .010 away from the cutting tool
    (B) .020 toward the cutting tool   (D) .020 away from the cutting tool

continued on next page

The steps listed below are for aligning the lathe centers by the trial-cut method. These steps *are not listed in order.* **In the right-hand column place the letter for each step beside the numbers 11 to 20 so that the steps will be in the proper sequence. That is, number 11 will be the first step and the last step will be number 20.**

A   Take a light trial cut for about .25 in. at the tailstock end of the work.   11.   _____

B   Move the carriage to the left until the cutting tool is about 1 in. from the lathe dog and   12.   _____
start the lathe.

C   Stop the lathe and measure both diameters with a micrometer.   13.   _____

D   Disengage the automatic feed and note the reading on the graduated collar.   14.   _____

E   Cut the left section about .50 in. long.   15.   _____

F   Turn the crossfeed handle clockwise until the graduated-feed collar reading is the same as   16.   _____
for the right-hand cut.

G   Turn the crossfeed handle counterclockwise to bring the cutting tool away from the work.   17.   _____

H   Continue to adjust the tailstock and take trial cuts until both diameters are the same.   18.   _____

I   Take another light cut from both ends at the same graduated collar setting and measure   19.   _____
both diameters.

J   If both diameters are not the same, adjust the tailstock one-half the difference in readings.   20.   _____

SCORE   _____

20

TEST 46   Facing Between Centers *(Unit 51)*

After the workpiece has been cut off, it is generally faced to provide a flat, square surface from which measurements may be taken. Facing may be done between centers but is more often done in a chuck.

Decide whether each statement is true or false and circle the letter in the right-hand column that indicates your choice.

1. Workpieces are generally cut off on the saw to the finished length and then faced in the lathe to square the ends.   1.   T   F

2. The toolpost should be set on the left side of the compound rest when facing.   2.   T   F

3. The facing tool should be set slightly below center.   3.   T   F

4. When facing is being done between centers, a quarter-center is often used.   4.   T   F

5. The point of the facing cutting tool should point slightly to the left.   5.   T   F

6. When facing, set the depth of cut by turning the apron handwheel.   6.   T   F

7. After the depth of cut has been set with the apron handwheel, the carriage lock must be tightened.   7.   T   F

8. The cutting tool should always be fed from the center to the outside of the workpiece.   8.   T   F

9. The lathe center must always be in line to produce a flat, square surface.   9.   T   F

10. When the compound rest is set to 30° for facing to length, .020 in. can be removed from the end of the work by feeding in .010 in. to the compound rest.   10.   T   F

SCORE _____

10

TEST 47   Machining Between Centers *(Unit 52)*

When work is being machined on a lathe, it is important that it be machined in as short a time as possible. It is, therefore, necessary that the student realizes the importance of efficient roughing and finishing cuts. When diameters of different sizes are turned on a workpiece a shoulder will result. Shoulders may be square, beveled, or filleted. The latter two types are stronger than the square shoulder and are used where strength is important.

Place the correct word(s) in the blank space(s) provided at the right-hand side of the page that will make the sentence complete and true.

1. Whenever possible, a piece of rough stock should be finish-turned to size in __?__ cuts.   1. _____

2. The first cut taken on the end of a workpiece to establish a size is called a __?__ cut.   2. _____

3. If the diameter of a workpiece is to be reduced from 1.800 to 1.760 in., the crossfeed handle must be turned in a distance of __?__ in.   3. _____

4. A roughing cut should bring the work to within __?__ of the finished size.   4. _____

5. A finish cut is used to bring the work to an accurate __?__ and to produce a good surface __?__ .   5. _____
   _____

6. When taking a roughing cut, never point the cutting tool to the __?__ .   6. _____

7. The lathe must always be __?__ when measuring work.   7. _____

8. The location of a shoulder may be marked with a center punch or by cutting a light __?__ around the work.   8. _____

9. A square shoulder should be rough-turned within __?__ in. of the required length.   9. _____

10. A __?__ tool should be used for machining a square shoulder.   10. _____

11. When turning a square shoulder, in order not to damage the small diameter, note the reading on the crossfeed-screw __?__ collar.   11. _____

12. When machining a filleted shoulder, turn the small diameter to the correct length minus the __?__ to be cut.   12. _____

13. When a filleted shoulder is being cut, the lathe speed should be one- __?__ the turning speed.   13. _____

14. A __?__ cutting tool is used to produce the correct form on a filleted shoulder.   14. _____

15. A __?__ cutting tool is used when machining a beveled shoulder.   15. _____

16. The cutting edge of this tool is set to the required angle with a __?__ .   16. _____

17. If chatter occurs when a beveled shoulder is being machined, it may be necessary to cut the bevel using the __?__ __?__ feed handle.   17. _____
   _____

18. To produce a good surface finish when cutting a filleted or beveled shoulder, use cutting __?__ .   18. _____

SCORE _____

20

TEST 48   Filing, Polishing, *(Units 52, 53)*
and Knurling

Filing and polishing are operations often performed on the lathe to remove very small amounts from the diameter and to improve the surface finish of the work. Knurling is used to improve the appearance of a workpiece and provide a good grip.

Decide whether each statement is true or false and circle the letter in the right-hand column that indicates your choice.

1.	It is considered good practice to file a diameter to size.	1.	T F
2.	Excessive filing will produce out-of-round work.	2.	T F
3.	The maximum amount to be left for filing should be .002 to .003 in.	3.	T F
4.	The spindle speed for filing is twice that for turning.	4.	T F
5.	Use short, quick strokes when filing in a lathe.	5.	T F
6.	It is good filing practice to grasp the file handle in the left hand.	6.	T F
7.	Clean the file frequently by tapping it on the lathe bed.	7.	T F
8.	A little chalk rubbed on the file will help prevent clogging.	8.	T F
9.	A mill file may be used for filing in a lathe.	9.	T F
10.	Always cover the lathe bed with a cloth when filing.	10.	T F
11.	Never use a file without a handle.	11.	T F
12.	When work is being polished in a lathe, the lead screw and the feed rod should be disengaged.	12.	T F
13.	To produce a fine surface finish when polishing, use 80- to 90-grit abrasive cloth.	13.	T F
14.	Oil should not be used when finish polishing the work.	14.	T F
15.	Cover the lathe bed with a piece of paper when polishing the work.	15.	T F
16.	*Knurling* is the process of impressing a rectangular-shaped pattern on the workpiece.	16.	T F
17.	Most knurling tools have self-centering heads.	17.	T F
18.	Knurling is performed at one-quarter the speed of turning.	18.	T F
19.	The knurling tool should be set up to point slightly to the left.	19.	T F
20.	The automatic feed may be disengaged at any time without damaging the knurl.	20.	T F

SCORE _____

20

TEST 49   Form Turning *(Unit 53)*

Form turning is used when it may be necessary to duplicate curved or irregular shapes on several workpieces.

Place the correct word(s) in the blank space(s) provided at the right-hand side of the page that will make the sentence complete and true.

1. Four methods of form turning are __?__ , __?__ , __?__ , and __?__ .

2. A freehand radius is machined by turning the carriage and the __?__ handle.

3. When a freehand radius is being cut, it is advisable to start at the __?__ diameter.

4. When only a few duplicate pieces are required and no tracer attachment is available, the parts can be made by fastening a __?__ to the back of the lathe and a follower on the __?__ __?__ of the lathe.

5. When the method outlined in question 4 is used, the cross slide must be __?__ from the crossfeed screw.

6. The workpiece is first roughed out close to form and size by __?__ .

7. The __?__ feed is used to move the cutting tool along for the finish cut.

8. The finish cut should not be any deeper than __?__ in.

9. Tracer lathes incorporate a means of moving the __?__ __?__ hydraulically.

10. Duplication of a part is achieved by a __?__ bearing against a __?__ of the desired profile.

11. When a hydraulic tracer lathe is used, the cutting tool point and the __?__ should have the form and radius.

12. The centerline of the template must be __?__ to the ways of the lathe.

13. No angle of larger than __?__ ° should be incorporated into the template form.

1. _____
   _____
   _____
   _____

2. _____

3. _____

4. _____
   _____
   _____

5. _____

6. _____

7. _____

8. _____

9. _____
   _____

10. _____
    _____

11. _____

12. _____

13. _____

SCORE _____

20

TEST 50   Tapers and Taper *(Unit 54)* Calculations

Tapers are very important in the machine trade. Standard tapers are used extensively to align and hold machine parts. A good machinist must be able to measure, calculate, and machine any taper accurately.

Select the most appropriate answer for each question and circle the letter in the right-hand column that indicates your choice.

1. Tapers may be expressed in
   (A) taper per foot
   (B) taper per inch
   (C) degrees
   (D) all of these

   1.   A   B   C   D

2. Metric tapers are expressed as
   (A) ratios
   (B) fractions
   (C) whole numbers
   (D) decimals

   2.   A   B   C   D

3. Which of the following does a taper *not* provide?
   (A) rapid alignment
   (B) an easy method of holding tools
   (C) accurate alignment
   (D) positive drive under heavy loads

   3.   A   B   C   D

4. A self-releasing, or steep, taper requires the use of a
   (A) draw bar
   (B) key
   (C) key and draw bar
   (D) setscrew

   4.   A   B   C   D

5. Which of the following is a self-releasing, or steep, taper?
   (A) Morse
   (B) Brown and Sharpe
   (C) Standard milling machine
   (D) Jarno

   5.   A   B   C   D

6. A self-holding taper has a
   (A) large taper angle
   (B) small taper angle
   (C) draw bar
   (D) key

   6.   A   B   C   D

7. The Morse taper has a taper per foot of
   (A) .625 in.
   (B) .750 in.
   (C) 3.500 in.
   (D) approximately .625 in.

   7.   A   B   C   D

8. The Jarno taper has a taper per foot of
   (A) .500 in.
   (B) .600 in.
   (C) .625 in.
   (D) .750 in.

   8.   A   B   C   D

9. The taper of most drills and machine reamers is
   (A) Brown and Sharpe
   (B) Morse
   (C) Jarno
   (D) Standard taper pin

   9.   A   B   C   D

10. The Standard taper pin has a taper per foot of
    (A) .250 in.
    (B) .625 in.
    (C) .500 in.
    (D) .625 in.

    10.   A   B   C   D

11. To calculate the taper per foot, which of the following is *not* required?
    (A) length of taper
    (B) length of work
    (C) large diameter
    (D) small diameter

    11.   A   B   C   D

12. The large diameter of a taper is 2 in., the small diameter is 1 in., the taper length is 4 in. What is the taper per foot?
    (A) 4 in.
    (B) 3 in.
    (C) 2 in.
    (D) 1 in.

    12.   A   B   C   D

13. In question 12, the taper per inch would be
    (A) .125 in.
    (B) .200 in.
    (C) .250 in.
    (D) .375 in.

    13.   A   B   C   D

continued on next page

14. The large diameter of a taper is 2.500 in., the small diameter is 2.000 in., and the length of taper is 10 in. What is the taper per inch?

(A) .025 in.   (C) .250 in.
(B) .050 in.   (D) .031 in.

14.   A   B   C   D

15. In question 14, the taper per foot would be

(A) .200   (C) .450
(B) .300   (D) .600

15.   A   B   C   D

16. If a metric taper is 1:15, this means that

(A) the work tapers 15 mm in 1 m   (C) the work tapers 1 mm in 15 mm
(B) the small diameter is 15 mm and the taper is 1 mm   (D) the large diameter is 15 mm and the taper is 1 mm

16.   A   B   C   D

17. The amount of taper per unit of length K is

(A) $1/K$ (C) $K/2$
(B) $2/K$ (D) $1 + K$

17.   A   B   C   D

18. The total taper on a workpiece equals

(A) taper per mm $\times$ length of work   (C) total taper $\times$ length of work
(B) taper per mm $\times$ length of taper   (D) total taper $\times$ length of taper

18.   A   B   C   D

19. The large diameter of a metric taper is equal to

(A) $\dfrac{\text{small diameter } + \text{ length of taper}}{\text{length of work}}$ (C) $\dfrac{\text{small diameter } + \text{ taper per unit length}}{\text{length of taper}}$

(B) $\dfrac{\text{small diameter } + \text{ length of work}}{\text{length of taper}}$ (D) $\dfrac{\text{small diameter } + \text{ length of taper}}{\text{taper per unit length}}$

19.   A   B   C   D

20. A workpiece has a taper of 1:25. The small diameter is 80 mm, the length of the taper is 125 mm, and the large diameter is

(A) 85 mm   (C) 90 mm
(B) 87.5 mm   (D) 92.5 mm

20.   A   B   C   D

SCORE _____

20

TEST 51   Taper Turning *(Unit 54)*

To turn a properly fitting taper, the machinist must take care to accurately set up the machine. The method used to turn the taper will depend on the work length, the taper length, the type of taper required, and the number of pieces to be machined.

Place the correct word(s) in the blank space(s) provided at the right-hand side of the page that will make the sentence complete and true.

1. Tapers may be produced on a lathe by three methods: using the taper __?__ , offsetting the __?__ , and swiveling the __?__ rest.

1. _____

_____

_____

2. Most taper attachments can be set to the taper per foot or to the __?__ of the workpiece.

2. _____

3. When cutting a taper using a taper attachment, set the __?__ bar to the required taper per foot or to the required __?__ of the taper.

3. _____

_____

4. When a taper is being turned, it is important to set the cutting tool on __?__ .

4. _____

5. Before starting a new cut when using the taper attachment, move the cutting tool .50 in. past the end of the work to remove the __?__ .

5. _____

6. When a plain taper attachment is being used, the depth of cut is set with the __?__ __?__ feed handle.

6. _____

_____

7. The compound rest is often used to cut short, __?__ tapers, which are shown on the print in __?__ .

7. _____

_____

8. The compound rest must be set to one- __?__ the included angle of the taper.

8. _____

9. External tapers can be checked for accuracy by using a __?__ ring __?__ .

9. _____

_____

10. When fitting a taper, draw __?__ equally spaced chalk lines along the length of the taper.

10. _____

11. When a taper is being checked with a micrometer, __?__ lines that are exactly 1 __?__ apart are required on the taper.

11. _____

_____

12. When a taper is being checked with a gage, the work should be turned in a __?__ direction for about one- __?__ a turn.

12. _____

_____

SCORE _____

20

TEST 52   Thread Types, Definitions, *(Unit 55)* and Calculations

Threads are one of the most widely used means of fastening parts together. Threads are produced by various methods in industry. In a machine shop, threads are generally cut on a lathe. Before cutting a thread, the machinist must have a knowledge of the various thread types, standards, and calculations.

Select the most appropriate answer for each question and circle the letter in the right-hand column that indicates your choice.

1. For which of the following is a thread *not generally* used?   1.   A   B   C   D
   (A) to permanently fasten two parts   (C) to provide a means of accurate
        together   measurement
   (B) to transmit motion   (D) to increase torque

2. Which of the following thread forms does *not* have a 60° angle?   2.   A   B   C   D
   (A) American National   (C) International Metric
   (B) American National Acme   (D) Unified

3. With which of the following threads may a Unified thread of the same pitch and diameter   3.   A   B   C   D
   be interchanged?
   (A) Acme   (C) International Metric
   (B) Square   (D) Whitworth

4. ISO metric threads were developed because of   4.   A   B   C   D
   (A) lack of international standards   (C) ease of calculation
   (B) ease of cutting   (D) all of these

5. The Unified thread was developed from a combination of the   5.   A   B   C   D
   (A) Square and Acme   (C) metric and Whitworth
   (B) American National and Whitworth   (D) American National and metric

6. Which of the following threads is best suited for feed screws, jacks, and vises?   6.   A   B   C   D
   (A) American National   (C) Unified
   (B) International Metric   (D) Acme

7. The angle of the American National thread is   7.   A   B   C   D
   (A) 29°   (C) 60°
   (B) 59°   (D) 30°

8. The distance a screw thread advances axially in one revolution is   8.   A   B   C   D
   (A) the lead   (C) the pitch
   (B) the thickness of the thread   (D) half the thickness of the thread

9. For American National Form threads, the pitch is equal to   9.   A   B   C   D
   (A) $\dfrac{1}{N}$ (C) $\dfrac{N}{D}$

   (B) $\dfrac{1}{D}$ (D) $\dfrac{1}{\frac{1}{2}D}$

10. The main difference between the National Form thread and the ISO metric thread is the   10.   A   B   C   D
    (A) included angle   (C) depth of external thread
    (B) width of crest   (D) all of these

11. How many thread sizes are in the ISO thread series?   11.   A   B   C   D
    (A) 15   (C) 40
    (B) 49   (D) 25

12. The depth of an external American National thread is   12.   A   B   C   D
    (A) .866 × pitch   (C) .6134 × pitch
    (B) .6495 × pitch   (D) .500 × pitch

continued on next page

13. The width of the crest for an external Unified thread is

    (A) .3707 × pitch   (C) .250 × pitch

    (B) .335 × pitch   (D) .125 × pitch

13.   A   B   C   D

14. The depth of a 3/8-in.—16 American National Standard thread would be

    (A) .383 in.   (C) .625 in.

    (B) .0383 in.   (D) .0312 in.

14.   A   B   C   D

15. The minor diameter of the thread in question 14 would be

    (A) .3298 in.   (C) .2983 in.

    (B) .3125 in.   (D) .375 in.

15.   A   B   C   D

16. The distance across the *flat* of a 3/4-in.—10 American National thread would be

    (A) .0125 in.   (C) .0625 in.

    (B) .125 in.   (D) .250 in.

16.   A   B   C   D

17. A typical metric thread is designated M14 × 2. The pitch of this thread is

    (A) 14 mm   (C) 2 mm

    (B) 7 mm   (D) 1 mm

17.   A   B   C   D

18. The depth of the thread in question 17 is

    (A) 1.22 mm   (C) 1.29 mm

    (B) 0.61 mm   (D) 0.64 mm

18.   A   B   C   D

19. The minor diameter of the thread in question 17 is

    (A) 12.65 mm   (C) 12.77 mm

    (B) 11.40 mm   (D) 11.54 mm

19.   A   B   C   D

20. The width of the crest of the thread in question 17 is

    (A) 0.12 mm   (C) 0.25 mm

    (B) 0.15 mm   (D) 0.27 mm

20.   A   B   C   D

SCORE _____

20

TEST 53   Thread Cutting *(Unit 55)*

Thread cutting on a lathe is an operation that requires the constant attention of the operator. In order to produce an accurate thread, the operator must set up the lathe, the cutting tool, and the workpiece properly.

Place the correct word(s) in the blank space(s) provided at the right-hand side of the page that will make the sentence complete and true.

1. The chasing dial indicates when the split nut should be engaged with the __?__ .

2. Engage the split-nut lever at any graduation on the dial for __?__ -number threads.

3. Engage the split-nut lever at any main division on the dial for __?__ -number threads.

4. Engage the split-nut lever at every other main division on the dial for __?__ threads.

5. Before a thread is cut, the end of the work should be __?__ .

6. Always mark the slot of the __?__ plate in which the tail of the dog is engaged.

7. Lathe speed for threading is __?__ the speed used for turning.

8. The compound rest should be set to the __?__ for cutting a left-hand thread.

9. The threading cutting tool is set at right angles to the work by means of a __?__ gage.

10. It is considered good practice to have the major diameter __?__ in. undersize when a thread is being cut in a lathe.

11. The first or trial cut, in threading, should be about __?__ in. deep.

12. The number of threads per inch is checked with a thread __?__ gage.

13. The depth of cut, in threading, is set by the __?__ -rest handle.

14. When a threading tool is being reset, the __?__ -nut lever must be engaged before the cutting tool is positioned in the groove.

15. After resetting a threading tool, set the __?__ graduated collar to __?__ .

16. For cutting an American National Acme thread, the compound rest should be set at __?__ °.

17. For cutting an *internal* left-hand American National thread, the compound rest should be swung to the __?__ .

18. Metric threads may be cut on a standard quick-change gear lathe by using two change gears having __?__ and __?__ teeth in the gear train.

1. _____
2. _____
3. _____
4. _____
5. _____
6. _____
7. _____
8. _____
9. _____
10. _____
11. _____
12. _____
13. _____
14. _____
15. _____
    _____
16. _____
17. _____
18. _____
    _____

SCORE _____

20

TEST 54   Steady Rests, Follower *(Unit 56)* Rests, and Mandrels

Steady rests, follower rests, and mandrels are accessories that increase the versatility of the lathe. Steady and follower rests permit the machining of long slender workpieces, while a mandrel fitted through the hole of a workpiece permits machining the sides square and/or the outside diameter concentric with the hole.

Decide whether each statement is true or false and circle the letter in the right-hand column that indicates your choice.

1. A steady rest is used to prevent long slender work from springing when being machined in a lathe.   1.   T   F

2. A steady rest has two jaws to support the workpiece.   2.   T   F

3. The jaws of a steady rest are made of hardened steel.   3.   T   F

4. The steady rest is fastened to the lathe carriage.   4.   T   F

5. A steady rest can only be used on work that is supported between lathe centers.   5.   T   F

6. The workpiece must have a trued section to provide a bearing surface for the steady rest jaws.   6.   T   F

7. Cutting fluid is used to lubricate the bearing between the work and the steady rest jaws.   7.   T   F

8. When machining long work that is held in a chuck, first adjust the steady rest to the work near the chuck.   8.   T   F

9. The lubricant may be used to indicate contact between the work and the steady rest jaws.   9.   T   F

10. Before machining the work when using a steady rest, align the top and front of two turned diameters with a dial indicator.   10.   T   F

11. A cathead must be used if round work is to be held accurately in a steady rest.   11.   T   F

12. The follower rest is fastened to the lathe carriage.   12.   T   F

13. The follower rest supports the work on the top and rear sides to prevent it from springing.   13.   T   F

14. All three jaws of the follower rest must be adjusted evenly against the work.   14.   T   F

15. The cutting tool should be mounted just to the right of the follower rest jaws.   15.   T   F

16. A follower rest is particularly useful when long shafts are being threaded.   16.   T   F

17. Solid mandrels are hardened and ground parallel.   17.   T   F

18. An expansion mandrel uses a tapered solid mandrel and grooved sleeve to fit odd-sized holes in workpieces.   18.   T   F

19. A gang mandrel is tapered with a threaded end to hold identical workpieces on the mandrel.   19.   T   F

20. The projecting portion of a taper-shank mandrel may be threaded, parallel, or tapered.   20.   T   F

SCORE _____

20

TEST 55   Chucks and Chuck Work *(Units 57, 58)*

Much of the work machined in a lathe may be held in a three- or four-jaw chuck. Operations such as turning and threading outside diameters, as well as drilling, boring, reaming, and tapping holes, may be performed.

Place the correct word(s) in the blank space(s) provided at the right-hand side of the page that will make the sentence complete and true.

1. To help preserve the chuck accuracy, the workpiece and the chuck jaws should be __?__ .

2. The jaws of a three-jaw universal chuck are moved in or out by means of a __?__ thread.

3. When work is trued in a four-jaw chuck, a piece of __?__ may be used to mark the high spot.

4. The high spot may also be checked with a __?__ gage.

5. If the work has to be set up accurately, an __?__ should be used.

6. When turning with PCBN tools, the manufacturer's recommendations for __?__ , __?__ , and __?__ should be followed.

7. When cutting off work in a lathe, be sure the cutting-off tool blade projects beyond the holder __?__ the diameter of the work plus __?__ in. for clearance.

8. When cutting off work, move the parting tool back and __?__ slightly to prevent __?__ .

9. A hole to be drilled in a workpiece may be spotted with a __?__ drill.

10. When a tapered-shank drill is mounted in the tailstock, a lathe __?__ should be used to stop it from turning.

11. When drilling, you may support the end of the drill with the end of the __?__ to prevent it from wobbling.

12. Always __?__ the pressure on the feed as the drill breaks through the work.

13. __?__ is the operation of trueing a drilled or cored hole with a single-pointed cutting tool.

14. The reaming allowance for a .500-in. hole should be __?__ in.

15. Reaming should be performed at __?__ the speed of drilling.

16. Never turn the lathe spindle or reamer __?__ when reaming.

1. _____

2. _____

3. _____

4. _____

5. _____

6. _____

_____

_____

7. _____

_____

8. _____

_____

9. _____

10. _____

11. _____

12. _____

13. _____

14. _____

15. _____

16. _____

SCORE _____

20

Lathes, Accessories, and Cutting Speeds Review Test

(Section 11)

PART 1

Place the letter of each illustrated lathe part beside the proper name in the right-hand column.

1. tailstock spindle clamp _____
2. speed-change levers _____
3. automatic-feed lever _____
4. feed-reverse lever _____
5. compound-rest feed handle _____
6. crossfeed-screw handle _____
7. split-nut lever _____
8. leadscrew _____
9. cross-slide _____
10. feed rod _____

Place the letter of each illustrated lathe accessory or operation beside the proper name in the right-hand column.

K.

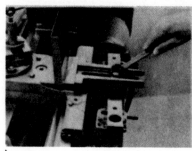

L.

M.   N.   O.

Q.

R.   S.

P.

11. type "L" spindle _____
12. carbide tool-holder _____
13. steady rest _____
14. left-hand cutoff tool _____
15. cathead _____
16. straight cutoff tool _____
17. cam-lock spindle _____
18. left-hand tool-holder _____
19. taper attachment _____
20. quick-change toolpost _____
21. follower rest _____
22. right-hand tool-holder _____
23. turret toolpost _____

T.

U.

V.

W.

PART 2

Select the most appropriate answer for each question and circle the letter in the right-hand column that indicates your choice.

24. The quick-change gearbox provides the drive for the   24.   A   B   C   D
    (A) leadscrew   (C) leadscrew and feed rod
    (B) feed rod   (D) lathe spindle

25. The apron handwheel usually moves the carriage by means of   25.   A   B   C   D
    (A) spur gears   (C) miter gears
    (B) bevel gears   (D) rack and pinion

26. Which of the following lathe parts is *not* attached to the carriage?   26.   A   B   C   D
    (A) saddle   (C) apron
    (B) quick-change gearbox   (D) cross slide

27. In order to machine work in the shortest time possible, you should   27.   A   B   C   D
    (A) increase the speed and decrease the feed (C) use the recommended speed and feed
    (B) decrease the speed and increase the feed (D) decrease the speed to increase the life of
                                                  the cutting tool

28. The r/min at which to set the lathe is calculated by which of the following formulas?   28.   A   B   C   D

    (A) $\dfrac{CS}{4 \times D}$ (C) $\dfrac{4 \times CS}{\pi D}$

    (B) $\dfrac{D}{4 \times CS}$ (D) $\dfrac{4 \times CS}{D}$

29. The r/min for rough turning a 3-in.-diameter machine-steel shaft (CS = 90) using a   29.   A   B   C   D
    high-speed cutting tool would be
    (A) 120   (C) 95
    (B) 100   (D) 90

30. *Feed* may be defined as the   30.   A   B   C   D
    (A) distance the cutting tool advances   (C) length of the cutting removed in one
         along the work for each revolution   revolution
    (B) length of the cutting removed in 1 min   (D) distance the cutting tool moves in 1 min

31. The taper in the headstock spindle should be cleaned by   31.   A   B   C   D
    (A) holding a cloth in the spindle while   (C) pushing a cloth into the hole with a
         the spindle is revolving   stick or rod
    (B) running your finger around the taper   (D) using an air hose

32. The fastest and most accurate way to align lathe centers is by   32.   A   B   C   D
    (A) using the tailstock graduations   (C) using a dial indicator and test bar
    (B) the trial-cut method   (D) eye

33. Tapers may be expressed in   33.   A   B   C   D
    (A) taper per foot   (C) degrees
    (B) taper per inch   (D) all of these

continued on next page

34. To calculate the taper per foot, which of the following is *not* required?   34.   A   B   C   D
  (A) length of taper   (C) large diameter
  (B) length of work   (D) small diameter

35. Which of the following thread forms does *not* have a 60° angle?   35.   A   B   C   D
  (A) American National   (C) International Metric
  (B) American National Acme   (D) Unified

PART 3

Place the correct word(s) in the blank space(s) provided at the right-hand side of the page that will make the sentence complete and true.

36. Always wear approved __?__ glasses when operating a lathe.   36.   _____

37. Do not take __?__ cuts on long slender workpieces.   37.   _____

38. Most taper attachments can be set to the taper per foot or to the __?__ of the workpiece.   38.   _____

39. Before a thread is cut, the end of the work should be __?__ .   39.   _____

40. The lathe speed for threading is one- __?__ that for turning.   40.   _____

41. The jaws of a three-jaw chuck are moved in and out by means of a __?__ thread.   41.   _____

42. When cutting off work in a lathe, be sure the cutting blade projects beyond the holder   42.   _____

  __?__ the diameter of the work plus __?__ in. for clearance.   _____

43. Never turn the lathe spindle or reamer __?__ when reaming.   43.   _____

PART 4

Decide whether each statement is true or false and circle the letter in the right-hand column that indicates your choice.

44. Set the toolpost to the left side of the compound rest when facing.   44.   T   F

45. When the compound rest is set to 30° for facing, .020 in. can be removed from the end   45.   T   F
  of the work by feeding in the compound rest .010 in.

46. Use short, quick strokes when filing in a lathe.   46.   T   F

47. The knurling tool should be set up to point slightly to the left.   47.   T   F

48. A steady rest is clamped to the ways of the lathe.   48.   T   F

49. Solid mandrels are hardened and ground parallel.   49.   T   F

                                                                     SCORE   _____
                                                                             50

TEST 56   The Vertical-Milling- *(Unit 59)*
Machine Setup—
Milling a Flat Surface

Much of the work done on a horizontal milling machine can be done faster on a vertical milling machine and often with less costly cutters. This machine can perform many operations such as face and end milling and keyway, dovetail, and T-slot cutting, as well as drilling, boring, and reaming. It is truly one of the most versatile machines in the machine shop.

Place the correct word(s) in the blank space(s) provided at the right-hand side of the page that will make the sentence complete and true.

1. If the vertical head is not aligned, drilled or bored holes will not be __?__ with the work.

2. When you are face milling and more than one pass is required, the machined surface will be __?__ if the head is not aligned.

3. When the head is being aligned, the indicator is mounted on a rod held in the __?__ or in a __?__ chuck.

4. The indicator is lowered until it touches the front of the table and registers about one- __?__ of a revolution.

5. Rotate the machine spindle __?__ and compare the readings on both sides of the table.

6. Swivel the head until the indicator registers one- __?__ the difference in the readings.

7. When the reading is the same on the back and front of the table, the head should be __?__ in this position.

8. To make the readings in the other direction, rotate the indicator __?__ °.

9. When the milling-machine vise is being aligned, the indicator is mounted in the __?__ .

10. When the vise is being aligned, the indicator bears against a __?__ held in the vise.

11. When aligning the vise, always tap it with a hammer so that the parallel moves in a direction __?__ __?__ the indicator.

12. When a narrow surface is being machined, the cutter should be __?__ than the surface.

13. Before taking a cut, be sure to tighten the __?__ clamp.

14. An angular surface may be cut by swiveling the __?__ to the correct angle or by setting the work in the vise at an __?__ .

15. When the work is set in the vise, the layout __?__ is set __?__ to the top of the vise.

1. _____

2. _____

3. _____
   _____

4. _____

5. _____

6. _____

7. _____

8. _____

9. _____

10. _____

11. _____
    _____

12. _____

13. _____

14. _____
    _____

15. _____
    _____

SCORE _____

19

TEST 57   Cutting Speeds, Feeds, and *(Unit 60)*
Depth of Cut

The life of the milling cutter and the efficiency of the cutting operation depend on the use of the correct cutting speeds, feeds, and depth of cut. A cutting speed that is too fast, a feed that is too great, and a cut that is too deep will shorten the life of a milling cutter. However, if a cutter is run too slowly, valuable time will be wasted. Therefore, the calculation and setting of the proper speeds, feeds, and depth of cut are important factors that an apprentice must learn.

Place the correct word(s) in the blank space(s) provided at the right-hand side of the page that will make the sentence complete and true.

1. *Cutting speed* is the rate in __?__ or __?__ per minute that a point on the circumference of a cutter should travel in 1 min.

   1. _____

   _____

2. The cutting speed of a metal will vary depending on the metal's hardness, structure, and __?__ .

   2. _____

3. The milling cutter must be set to revolve at a specified number of r/min depending on its __?__ .

   3. _____

4. In order to calculate the number of r/min to set a cutter, you must know the cutting speed of the __?__ and the __?__ of the cutter.

   4. _____

   _____

5. The formula for calculating the r/min for inch cutters is __?__ times the cutting speed divided by the __?__ of the cutter.

   5. _____

   _____

6. The formula for calculating the r/min for metric cutters is __?__ times the cutting speed divided by the __?__ of the cutter.

   6. _____

   _____

7. A 2-in.-diameter high-speed steel cutter should revolve at __?__ r/min to cut machine steel (100 CS) efficiently.

   7. _____

8. A 75-mm-diameter carbide cutter should revolve at __?__ r/min to cut tool steel (50 CS) efficiently.

   8. _____

9. *Feed,* measured in inches or millimeters per minute, is the rate at which the __?__ moves into the revolving __?__ .

   9. _____

   _____

10. The amount of material that should be removed by each tooth of the cutter is called __?__ per tooth.

    10. _____

11. Milling feed, in inches or meters per minute, is calculated by multiplying the number of teeth in the cutter by the __?__ per __?__ and the __?__ of the cutter.

    11. _____

    _____

    _____

12. Roughing cuts should be taken with __?__ cutters having fewer teeth.

    12. _____

13. A fine surface finish is produced when the feed is __?__ rather than speeding up the cutter.

    13. _____

SCORE _____

20

TEST 58   Machining a Block Square *(Unit 60)*
and Parallel–Cutting a
Keyseat

When a block is being machined square and parallel, there is a definite order in which the sides must be machined. Using the illustration below, place the letter that indicates each side beside the proper numbered side in the right-hand column.

1.   side #1 _____

side #2 _____

side #3 _____

side #4 _____

Place the correct word(s) in the blank space(s) provided at the right-hand side of the page that will make the sentence complete and true.

2. Remove all __?__ from the workpiece before setting it up.

2. _____

3. In order for the work to be seated properly on the parallels, __?__ feelers should be placed under each corner and the work should be "seated" with a __?__ hammer.

3. _____
_____

4. When machining the second side, place a __?__ bar between the side of the work and the __?__ jaw of the vise.

4. _____
_____

5. When positioning the bar (question 4), place it in about the __?__ of the amount of work held inside the vise.

5. _____

6. When the ends of the block are being machined, short pieces are usually held __?__ and are aligned with a __?__ .

6. _____
_____

7. Longer pieces are held __?__ and the end is machined with the __?__ of the end mill.

7. _____
_____

8. When laying out a keyseat, reference or center lines are first scribed on the __?__ of the shaft.

8. _____

9. Long shafts may be mounted in a table __?__ or in V-blocks.

9. _____

10. If the keyseat has two blind ends, a __?__ - or __?__ -fluted end mill should be used.

10. _____
_____

11. When centering a workpiece, place __?__ between the edge of the work and the cutter.

11. _____

12. The depth of a keyseat should be one- __?__ the width of the key.

12. _____

SCORE _____

20

TEST 59 Horizontal Milling Machine *(Unit 64)*
Parts and Accessories

One of the most versatile machines found in a machine shop is the horizontal milling machine. This machine tool can be used for flat- and form-milling operations, gear cutting, cam milling, drilling, boring, reaming, and a variety of other operations. Most machine tools have certain main operative parts, and it is important that the student know their location and purpose in order to operate the milling machine properly. A wide variety of attachments and accessories greatly increase the types of operations that can be performed on a horizontal milling machine.

PART 1

Place the letter of each illustrated horizontal milling machine part beside the proper name in the right-hand column.

1. base _____
2. column _____
3. knee _____
4. saddle _____
5. table _____
6. crossfeed handle _____
7. table handwheel _____
8. spindle _____
9. overarm _____
10. arbor support _____
11. vertical feed crank _____
12. knee clamp _____
13. table feed lever _____

PART 2

Place the name of each illustrated milling-machine attachment or accessory in the proper space in the right-hand column.

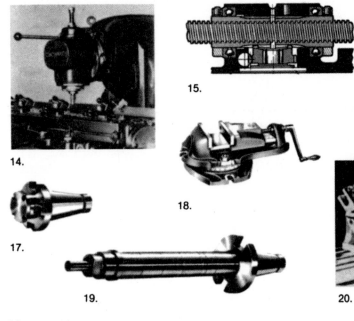

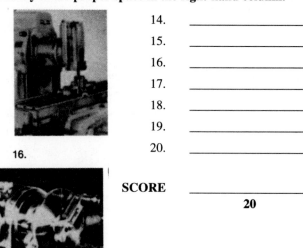

14. _____

14. _____
15. _____
16. _____
17. _____
18. _____
19. _____
20. _____

SCORE _____

20

15.

16.

17.

18.

19.

20.

TEST 60 Milling Cutters *(Unit 65)*

A *milling cutter* is a rotary cutting tool having a cylindrical body and equally spaced teeth around its periphery. As the cutter rotates on its axis, the teeth engage with the workpiece and remove metal in the form of chips. Milling cutters are available in a wide assortment of sizes and shapes to suit a variety of milling operations required on a workpiece. The apprentice or machine operator must be able to select the proper cutter for each job in order to produce accurate work and achieve the best results on a milling machine.

PART 1

Place the number of each illustrated cutter beside the proper name in the right-hand column.

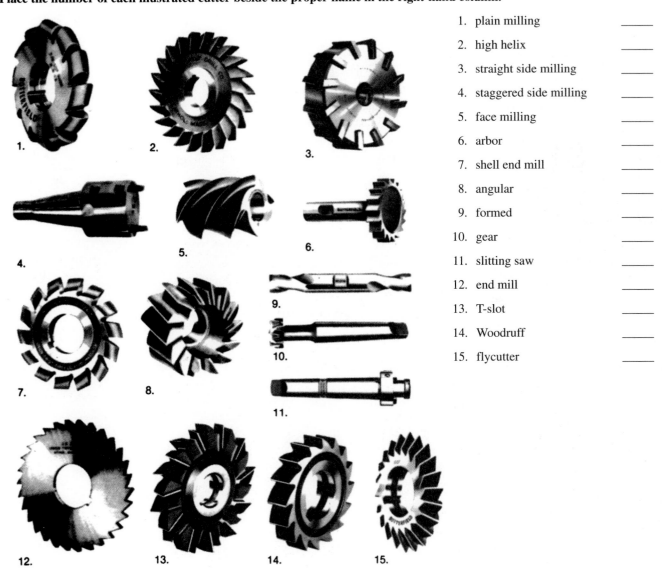

1. plain milling _____
2. high helix _____
3. straight side milling _____
4. staggered side milling _____
5. face milling _____
6. arbor _____
7. shell end mill _____
8. angular _____
9. formed _____
10. gear _____
11. slitting saw _____
12. end mill _____
13. T-slot _____
14. Woodruff _____
15. flycutter _____

PART 2

Select the most appropriate answer for each question and circle the letter in the right-hand column that indicates your choice.

16. The plain milling cutter is used to produce a(n)   16.   A   B   C   D
    (A) flat surface parallel to its axis   (C) vertical surface at 90° to its axis
    (B) contoured surface   (D) angular surface

17. Side milling cutters are generally used for   17.   A   B   C   D
    (A) cutting slots   (C) neither A nor B
    (B) face and straddle milling   (D) both A and B

18. Which of the following is *not* a formed cutter?   18.   A   B   C   D
    (A) concave   (C) helical
    (B) gear tooth   (D) convex

19. When both angles on a double-angle milling cutter are not the same, the cutter is designated by the   19.   A   B   C   D
    (A) left-side angle   (C) major angle
    (B) right-side angle   (D) included angle

20. When a two-fluted end mill is used to cut a slot, the maximum depth of cut should be   20.   A   B   C   D
    (A) twice the cutter diameter   (C) one-quarter the cutter diameter
    (B) one-half the cutter diameter   (D) the same as the cutter diameter

SCORE   _____

20

TEST 61   Milling-Machine Setups *(Units 66, 67, & 68)* and Indexing

The milling machine is a very sensitive machine capable of performing many operations other than milling, such as drilling, boring, reaming, and gear cutting. It is important that the machine is set up properly and that the operator is able to assess and correct any problem as it arises. A careless attitude on this machine can result in personal injury or damage to the machine and the workpiece.

Select the most appropriate answer for each question and circle the letter in the right-hand column that indicates your choice.

1. Which of the following statements is *not* considered good safety practice for an operator working on a milling machine?
   - (A) Be sure that the cutter and arbor clear the work.
   - (B) Adjust the work only when the cutter is stopped.
   - (C) Use a cloth to clean away the cuttings.
   - (D) Never reach over a revolving cutter.

   1.   A   B   C   D

2. When removing a milling cutter, always
   - (A) handle it with a cloth
   - (B) place it on the machine table
   - (C) place it on the floor
   - (D) any of these

   2.   A   B   C   D

3. When mounting a slitting saw,
   - (A) key the saw and tighten the arbor nut by hand
   - (B) key the saw and tighten the arbor nut securely
   - (C) do not key the saw but tighten the arbor nut securely
   - (D) oil the saw and arbor

   3.   A   B   C   D

4. When the arbor nut is tightened, it should be
   - (A) tightened with a wrench and soft hammer
   - (B) hand tightened with a wrench
   - (C) run on by power
   - (D) hand tightened

   4.   A   B   C   D

5. The arbor is held securely in the spindle by means of
   - (A) a steep taper
   - (B) a self-holding taper
   - (C) a draw-in bar
   - (D) both B and C

   5.   A   B   C   D

6. When the table is being aligned, the indicator should be mounted on the
   - (A) column
   - (B) arbor
   - (C) knee
   - (D) table

   6.   A   B   C   D

7. When a vise is being aligned, the indicator may *not* be mounted on the
   - (A) cutter
   - (B) face of the column
   - (C) solid vise jaw or parallel in the vise
   - (D) arbor

   7.   A   B   C   D

8. When aligning the vise, always tap the vise so that the end against which the indicator is bearing
   - (A) moves away from the indicator for one-half the amount of misalignment
   - (B) moves toward the indicator for one-half the amount of misalignment
   - (C) moves away from the indicator for the full amount of misalignment
   - (D) moves toward the indicator for the full amount of misalignment

   8.   A   B   C   D

9. To ensure that the workpiece to be milled is sitting firmly on the parallels in the vise,
   - (A) check each corner with a .005-in. feeler gage
   - (B) place a paper strip under each corner
   - (C) set the work directly on parallels
   - (D) strike it hard on each corner with a ball-peen hammer

   9.   A   B   C   D

10. The milling-machine arbor is driven by the
   - (A) holding power of the taper
   - (B) holding power of the draw-in bar
   - (C) drive keys or lugs
   - (D) all of these

   10.   A   B   C   D

11. When a surface is cut parallel to the axis of the arbor, the procedure is called
   - (A) side milling
   - (B) end milling
   - (C) face milling
   - (D) plain milling

   11.   A   B   C   D

continued on next page

12. If an accurate depth of cut is required, raise the table until the revolving cutter just

(A) touches the work

(B) touches the work and then lower the table .002 in.

(C) cuts a piece of paper between the cutter and the work

(D) removes bluing from the work surface

12. A   B   C   D

13. When a fine finish is required,

(A) increase the cutter speed

(B) reduce the feed

(C) use more cutting fluid

(D) both A and B

13. A   B   C   D

14. Which of the following is *not* a part of the dividing-head set?

(A) headstock

(B) steady rest

(C) worm wheel

(D) footstock

14. A   B   C   D

15. The simplest form of indexing is

(A) direct

(B) simple

(C) angular

(D) differential

15. A   B   C   D

16. Which of the following is *not* required for direct indexing?

(A) sector arms

(B) index crank

(C) worm wheel

(D) all of these

16. A   B   C   D

17. Which of the following hole circles can be used to index for nine divisions?

(A) 21

(B) 23

(C) 27

(D) 29

17. A   B   C   D

18. How many complete turns of the index crank would be required to mill an octagon?

(A) 40

(B) 20

(C) 10

(D) 5

18. A   B   C   D

19. Which of the following shapes would be produced if the crank were turned 6 full turns plus 14 holes on the 21-hole circle?

(A) triangle

(B) square

(C) hexagon

(D) octagon

19. A   B   C   D

20. Indexing a workpiece for a 72° angle would require

(A) 8 complete turns

(B) 7 complete turns and 14 holes on the 18-hole circle

(C) 6 complete turns

(D) 7 complete turns and 15 holes on the 27-hole circle

20. A   B   C   D

SCORE _____

20

TEST 62   Helical, Cam, and Clutch Milling *(Units 69, 70)*

The versatility of the horizontal milling machine and the variety of attachments available make it possible to perform special operations such as helical, cam, and clutch milling. Helical gears are used to provide additional strength to the gear teeth and they operate more quietly. Cams and clutches are used in mechanical devices to provide a drive or to change motions.

Select the most appropriate answer for each question and circle the letter in the right-hand column that indicates your choice.

1. Which of the following is *not* required when the operator is setting up to cut a helix?　　　　1.　A　　B　　C　　D
   (A) gear the dividing head to the leadscrew　(C) rotate the work
   (B) disengage the worm shaft　(D) swing the table

2. A helix is produced on a　　　　2.　A　　B　　C　　D
   (A) cylinder　(C) plane
   (B) cone　(D) both B and C

3. The formula used to calculate the angle of a helix is: Tangent of the helix angle =　　　　3.　A　　B　　C　　D
   (A) $\dfrac{\text{lead of helix}}{\text{circumference of work}}$　(C) $\dfrac{\text{diameter of work}}{\text{lead of helix}}$
   (B) $\dfrac{\text{circumference of work}}{\text{lead of helix}}$　(D) $\dfrac{\text{lead of helix}}{\text{diameter of work}}$

4. To cut a right-hand helix, swivel the table of the machine　　　　4.　A　　B　　C　　D
   (A) to the right　(C) clockwise
   (B) to the left　(D) counterclockwise

5. To cut a lead of 10 in. on a cylinder of 1.989-in. diameter, swing the table to　　　　5.　A　　B　　C　　D
   (A) 28° (tan = .53171)　(C) 32° (tan = .62847)
   (B) 30° (tan = .57735)　(D) 36° (tan = .72654)

6. When gearing a milling machine to the dividing head to cut a helix, you should connect the　　　　6.　A　　B　　C　　D
   (A) leadscrew to the worm shaft　(C) worm shaft and the spindle
   (B) leadscrew to the spindle　(D) both B and C

7. The formula used to calculate the change gears required to cut a given helix is:　　　　7.　A　　B　　C　　D
   $\dfrac{\text{Driven gears}}{\text{Driver gears}} =$
   (A) $\dfrac{\text{lead of machine}}{\text{lead of helix}}$　(C) $\dfrac{\text{lead of machine}}{\text{circumference of work}}$
   (B) $\dfrac{\text{lead of helix}}{\text{lead of machine}}$　(D) $\dfrac{\text{diameter of work}}{\text{lead of machine}}$

8. To change the direction of rotation of the part in order to cut an opposite helix, you should　　　　8.　A　　B　　C　　D
   (A) reverse the machine　(C) reverse the feed
   (B) interchange the gears on the spindle and the leadscrew　(D) insert an idler into the gear train

9. Which of the following equations is correct in regard to helical gear formulae?　　　　9.　A　　B　　C　　D
   (A) DP = NDP × cos helix angle　(C) NDP = DP ÷ sec helix angle
   (B) NDP = DP × cos helix angle　(D) none of these

10. Continued fractions is　　　　10.　A　　B　　C　　D
   (A) a method of calculating change gears　(C) a method of calculating ratios
   (B) another name for successive quotients　(D) all of these

continued on next page

11. A cam *cannot* be used   11.   A   B   C   D
- (A) to convert rotary motion to reciprocating motion
- (C) to convert linear motion to reciprocating motion
- (B) to convert reciprocating motion to rotary motion
- (D) as a locking device

12. Which of the following is *not* a type of cam follower?   12.   A   B   C   D
- (A) roller
- (C) finger
- (B) plunger
- (D) knife-edge

13. Which of the following is *not* a cam motion?   13.   A   B   C   D
- (A) uniform
- (C) uniform-harmonic
- (B) harmonic
- (D) uniformly accelerated and decelerated

14. Which of the following is *not* part of a cam?   14.   A   B   C   D
- (A) pitch
- (C) rise
- (B) lead
- (D) lobe

15. In a positive-type cam, the follower is always   15.   A   B   C   D
- (A) controlled by a spring
- (C) gravity driven
- (B) controlled by the cam
- (D) none of these

16. The cam that imparts the smoothest motion to the follower is   16.   A   B   C   D
- (A) harmonic
- (C) uniform-harmonic
- (B) uniformly accelerated and decelerated
- (D) uniform

17. Compared with the lead for a single-lobe cam, the lead for a double-lobe cam is   17.   A   B   C   D
- (A) half
- (C) twice
- (B) the same, but two cuts are taken
- (D) the same, but the work is rotated 180°

18. The formula for the pitch of a rack is   18.   A   B   C   D

(A) $\dfrac{3.1416}{DP}$

(C) $\dfrac{CP}{3.1416}$

(B) $\dfrac{DP}{3.1416}$

(D) $DP \times 3.1416$

19. The shortest lead that can be cut with regular change gears on a milling machine is .670 in. In order to cut a shorter lead, you must   19.   A   B   C   D
- (A) compound the gears
- (C) use smaller gears
- (B) change the rise of the cam
- (D) swivel the work and the vertical head to an angle

20. Which of the following is *not* a standard type of clutch?   20.   A   B   C   D
- (A) straight tooth
- (C) inclined tooth
- (B) staggered tooth
- (D) saw tooth

SCORE _____

20

Milling-Machine Accessories (Section 12) and Operations Review Test

PART 1

Place the letter of each illustration beside the proper name in the right-hand column.

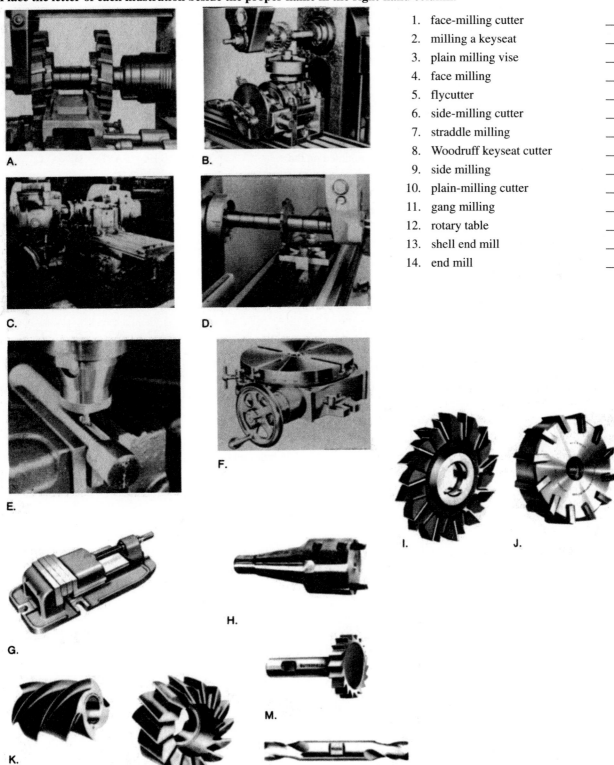

A.

B.

C.

D.

E.

F.

G.

H.

I.

J.

K.

L.

M.

N.

1. face-milling cutter _____
2. milling a keyseat _____
3. plain milling vise _____
4. face milling _____
5. flycutter _____
6. side-milling cutter _____
7. straddle milling _____
8. Woodruff keyseat cutter _____
9. side milling _____
10. plain-milling cutter _____
11. gang milling _____
12. rotary table _____
13. shell end mill _____
14. end mill _____

continued on next page

PART 2

Select the most appropriate answer for each question and circle the letter in the right-hand column that indicates your choice.

15. The plain milling cutter is used to produce a
  (A) flat surface parallel to its axis   (C) vertical surface at 90° to its axis
  (B) contoured surface   (D) angular surface

15.   A   B   C   D

16. Which of the following is *not* a formed cutter?
  (A) concave   (C) helical
  (B) gear tooth   (D) convex

16.   A   B   C   D

17. When removing a milling cutter, always
  (A) handle it with a cloth   (C) place it on the floor
  (B) place it on the machine table   (D) any of these

17.   A   B   C   D

18. When the table is being aligned, the indicator should be mounted on the
  (A) column   (C) knee
  (B) arbor   (D) table

18.   A   B   C   D

19. When the vise is being aligned, the indicator may *not* be mounted on the
  (A) cutter   (C) solid jaw or parallel in vise
  (B) face of the column   (D) arbor

19.   A   B   C   D

20. When a two-fluted end mill is used to cut a slot, the maximum depth of cut should be
  (A) twice the cutter diameter   (C) one-quarter the cutter diameter
  (B) one-half the cutter diameter   (D) the same as the cutter diameter

20.   A   B   C   D

21. Which of the following shapes would be produced if the crank were turned 13 full turns plus 7 holes on the 21-hole circle?
  (A) octagon   (C) square
  (B) hexagon   (D) triangle

21.   A   B   C   D

22. Which of the following is *not* required for direct indexing?
  (A) sector arms   (C) worm wheel
  (B) index crank   (D) all of these

22.   A   B   C   D

23. Which of the following hole circles can be used to index for 11 divisions?
  (A) 21   (C) 33
  (B) 27   (D) 39

23.   A   B   C   D

24. Indexing for a workpiece for a 60° angle would require
  (A) 6 complete turns   (C) 6 complete turns and 26 holes on the 39-
  (B) 7 complete turns and 14 holes on the   hole circle
     18-hole circle   (D) 7 complete turns

24.   A   B   C   D

PART 3

Place the correct word(s) in the blank space(s) provided at the right-hand side of the page that will make the sentence complete and true.

25. A 3.500-in.-diameter high-speed steel cutter should revolve at __?__ r/min to cut aluminum (700 CS) efficiently.

25. _____

26. The distance from the bottom of the tooth to the pitch circle is called the __?__ .

26. _____

27. When setting up to cut a gear, place the __?__ end of the mandrel on the indexing-head center.

27. _____

28. When you are face milling and more than one pass is required, the machined surface will be __?__ if the head is not aligned.

28. _____

29. Before taking a cut on the vertical milling machine, be sure to tighten the __?__ clamp.

29. _____

30. Remove all __?__ before setting the workpiece in a vise.

30. _____

31. In order for the workpiece to be seated properly on parallels, in a vise, __?__ feelers should be placed under each corner.

31. _____

Using the information given, calculate the dimensions listed in the right-hand column and place your answer beside each.

32. A gear has a diametral pitch of 8 and a pitch diameter of 5 in.

32. number of teeth _____
     outside diameter _____
     whole depth   _____

PART 4

Decide whether each statement is true or false and circle the letter in the right-hand column that indicates your choice.

33. The cutting speed of a metal is stated in in./min.   33.   T   F

34. In order to get a fine surface finish, you should increase the speed of the cutter.   34.   T   F

35. Always use a brush to clean away cuttings.   35.   T   F

36. The milling-machine arbor has a self-holding taper.   36.   T   F

37. The *addendum* is the distance from the outside diameter to the pitch circle.   37.   T   F

38. The outside diameter of a gear is equal to the number of teeth plus two divided by the   38.   T   F
pitch diameter.

39. When the head is being aligned on a vertical milling machine, the indicator should be   39.   T   F
mounted on a rod held in a chuck.

40. When indicating the head, swivel it the amount of difference in the readings on the dial 40.   T   F
indicator taken at the front and back of the table.

41. When aligning the vise, tap it so that the parallel moves away from the indicator.   41.   T   F

42. When machining a keyseat having two blind ends, use a two-fluted end mill.   42.   T   F

SCORE _____

44

TEST 63   Abrasives and Grinding Wheels *(Unit 71)*

The widespread use of abrasives in industry has made possible the refined manufacturing processes that have given us such a high standard of living. Because of these hard, tough grains, grinding wheels and other abrasive products have been developed that produce the finer finishes and dimensional accuracy required for interchangeable manufacture.

Select the most appropriate answer for each question and circle the letter in the right-hand column that indicates your choice.

1. Which of the following is *not* a manufactured abrasive?   1.   A   B   C   D
   (A) silicon carbide   (C) emery
   (B) aluminum oxide   (D) boron carbide

2. The most commonly used grinding wheels are made from   2.   A   B   C   D
   (A) aluminum oxide   (C) silicon carbide
   (B) diamond   (D) boron carbide

3. Aluminum-oxide wheels are generally used for grinding   3.   A   B   C   D
   (A) brass   (C) high-tensile-strength materials
   (B) cast iron   (D) none of these

4. Silicon carbide is generally used to grind   4.   A   B   C   D
   (A) ceramics   (C) copper
   (B) cemented carbide   (D) all of these

5. Which of the following statements is *not true* of boron carbide?   5.   A   B   C   D
   (A) It is used in higher-quality grinding   (C) It is used in the manufacture of precision
       wheels.   gages.
   (B) It is a cheap substitute for diamond dust. (D) It is harder than silicon carbide.

6. The softest manufactured abrasive is   6.   A   B   C   D
   (A) silicon carbide   (C) diamond
   (B) aluminum oxide   (D) cubic boron nitride

7. The best-coated abrasive to use on aluminum would be   7.   A   B   C   D
   (A) garnet   (C) aluminum oxide
   (B) emery   (D) silicon carbide

8. Which of the following statements is *not true* regarding a grinding wheel?   8.   A   B   C   D
   (A) It must be able to produce new cutting   (C) It must be able to break down gradually.
       edges.   (D) It must be hard and tough.
   (B) It must be soft and tough.

continued on next page

Place the correct word(s) in the blank space(s) provided at the right-hand side of the page that will make the sentence complete and true.

9. The material components of a grinding wheel are the abrasive grain and the __?__ .

9. _____

10. The __?__ indicates the strength with which the bond holds the abrasive particles together.

10. _____

11. The __?__ indicates the density of the grinding wheel.

11. _____

12. An open-structure wheel provides greater __?__ clearance than does a denser wheel.

12. _____

13. The __?__ of a wheel can be controlled by mold pressure.

13. _____

14. A small grinding wheel may be tested for cracks by lightly __?__ it with the handle of a screwdriver.

14. _____

15. The information regarding a small or medium grinding wheel is printed on the __?__ on the sides of the wheel.

15. _____

Place the proper position number from the following grinding-wheel blotter beside the corresponding information in the right-hand column.

16.

32A	46	H	8	V
Position #	1	2	3	4	5

16. structure _____

grit size _____

abrasive type _____

grade _____

bond type _____

SCORE _____

20

TEST 64   Surface Grinders and *(Units 72, 73)*
Operations

The surface grinder is used mainly for the production of highly finished flat surfaces, although contoured and irregular surfaces can also be ground. The versatility of the surface grinder depends upon the accessories available and the skill of the operator.

Select the most appropriate answer for each question and circle the letter in the right-hand column that indicates your choice.

1. The most common surface grinder found in a machine shop has a   1.   A   B   C   D
    (A) vertical spindle and reciprocating table   (C) vertical spindle and rotary table
    (B) horizontal spindle and reciprocating   (D) horizontal spindle and rotary table
        table

2. Which of the following statements is *not true* regarding a vertical spindle grinder with a   2.   A   B   C   D
    rotary table?
    (A) It grinds on the wheelface.   (C) It grinds on the wheel periphery.
    (B) It is the most efficient form of surface   (D) The surface pattern appears as a series
        grinder.   of intersecting arcs.

3. The surface grinder that can take the heaviest cut has a   3.   A   B   C   D
    (A) horizontal spindle and reciprocating   (C) vertical spindle and reciprocating table
        table   (D) vertical spindle and rotary table
    (B) horizontal spindle and rotary table

4. When selecting a grinding wheel for a job, always choose a   4.   A   B   C   D
    (A) soft wheel for hard material   (C) silicon-carbide wheel for grinding
    (B) hard wheel for hard material   machine steel
                                                   (D) aluminum-oxide wheel for grinding
                                                       aluminum

5. A wheel will glaze if the   5.   A   B   C   D
    (A) grain is too large   (C) wheel speed is too slow
    (B) work speed is too fast   (D) wheel is too hard

6. If the wheel wears too quickly, the   6.   A   B   C   D
    (A) wheel speed is too fast   (C) wheel is too soft
    (B) work speed is too slow   (D) structure is too open

7. Which of the following statements is *not true* when a grinding wheel is being mounted?   7.   A   B   C   D
    (A) The wheel is not cracked.   (C) The spindle is clean.
    (B) There is a blotter on one side of the   (D) The wheel does not bind on the spindle.
        wheel.

8. When you are taking a roughing cut on a surface grinder (horizontal spindle, reciprocat-   8.   A   B   C   D
    ing table), the depth of cut should be
    (A) .001 to .003 in.   (C) .015 to .020 in.
    (B) .004 to .006 in.   (D) .020 to .030 in.

9. Thin work is held for surface grinding on   9.   A   B   C   D
    (A) a parallel   (C) a piece of paper
    (B) a chuck block   (D) an adapter plate

10. When the grinding wheel is set to the work, the wheel should overlap the edge of the work by   10. A   B   C   D
    (A) .06 in.   (C) .18 in.
    (B) .12 in.   (D) .25 in.

continued on next page

Place the correct word(s) in the blank space(s) provided at the right-hand side of the page that will make the sentence complete and true.

11. *Trueing* is the operation of removing __?__ spots on the wheel and making it __?__ with the spindle.

11. _____

_____

12. The operation of removing dull grains and metal particles from the grinding wheel is called __?__ .

12. _____

13. When dressing a grinding wheel, offset the diamond dresser about __?__ in. to the __?__ of the wheel centerline.

13. _____

_____

14. When mounting work on a magnetic chuck, place a piece of __?__ between the work and the chuck.

14. _____

15. A short workpiece must be supported by steel __?__ to prevent it from moving under the grinding force.

15. _____

16. Before grinding, it is important to remove all __?__ from the workpiece.

16. _____

17. Work is usually mounted on an angle plate if one side must be ground __?__ with another.

17. _____

18. The trueing devices used for CBN grinding wheels are impregnated diamond __?__ and brake-controlling trueing devices.

18. _____

SCORE _____

20

Grinders, Parts, and Accessories

(Units 72, 73, 74)

The versatility of any machine depends to a large extent on the accessories available. There are many accessories available for grinding machines; some are shown in the following illustrations. Some illustrations show the accessory, while others show an operation that you are required to identify.

Place the letter of each illustrated part beside the proper name in the right-hand column.

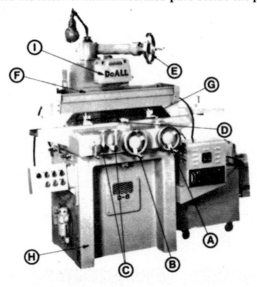

1. crossfeed handwheel _____

2. table reverse dogs _____

3. table control levers _____

4. table traverse handwheel _____

5. wheelfeed handwheel _____

6. wheelhead _____

7. table _____

8. base _____

9. table reverse lever _____

Place the letter of each illustrated part or operation beside the proper name in the right-hand column.

J.

K.

10. testing a grinding wheel _____

11. universal cylindrical grinder _____

12. principle of centerless grinding _____

13. balancing a grinding wheel _____

14. diamond dresser _____

15. magnetic V-block _____

16. magnetic chuck _____

17. magnetic chuck block _____

18. sine chuck _____

19. adjustable angle vise _____

20. toolpost grinder _____

SCORE _____

20

continued on next page

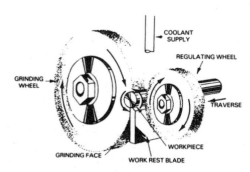

L.

M.

N.

O.

P.

Q.

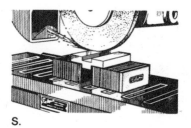

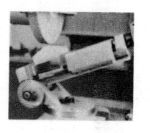

R.

S.

T.

TEST 66   Universal Cutter and *(Unit 75)*
Tool Grinder

The universal cutter and tool grinder is designed primarily to sharpen and renew all types of milling cutters, reamers, and taps. It may be used to a limited degree to perform such operations as cylindrical, taper, internal, and surface grinding. These latter operations cannot be performed as efficiently on this type of machine as on the proper type of grinding machine.

Place the letter of each illustrated part beside the proper name in the right-hand column.

1. footstock _____
2. wheelhead handwheel _____
3. lower table _____
4. headstock _____
5. infeed handwheel _____
6. upper table _____
7. saddle _____
8. stop dogs _____
9. wheelhead _____
10. table traverse crank _____

Place the letter of each illustrated accessory or operation beside the proper name in the right-hand column.

K.

11. measuring the helix angle _____
12. clearance grinding _____
13. hollow grinding _____
14. circle grinding _____
15. checking the cutter clearance angle _____
16. centering the wheelhead _____
17. centering the tooth rest _____
18. cutter clearance dial _____
19. cylindrical grinding _____
20. internal grinding _____

SCORE _____
20

continued on next page

L.

M.

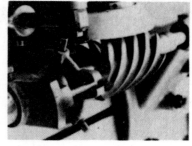

N.

O.

P.

Q.

R.

S.

T.

TEST 67 Milling Cutters *(Unit 75)*

To grind a milling cutter correctly, the operator must know the names and the function of the various cutter parts.

Place the letter of each illustrated cutter part beside the proper name in the right-hand column.

1. land _____
2. back of tooth _____
3. tooth face _____
4. secondary clearance _____
5. chip space _____
6. gullet _____
7. tooth angle _____
8. primary clearance _____
9. rake angle _____
10. heel _____

Select the most appropriate answer for each question and circle the letter in the right-hand column that indicates your choice.

11. The primary clearance ground on a high-speed steel milling cutter for cutting machine steel should be
    (A) 4°   (C) 8°
    (B) 6°   (D) 10°   11. A B C D

12. The best wheel to use for clearance grinding is a
    (A) flared cup (C) straight
    (B) diamond   (D) cutoff   12. A B C D

13. The clearance on a reamer is produced by
    (A) clearance grinding (C) circle grinding
    (B) hollow grinding   (D) centerless grinding   13. A B C D

14. The best wheel to use for hollow grinding is a
    (A) 4-in. flared cup   (C) 6-in. saucer
    (B) 4-in. straight cup (D) 6-in. cutoff   14. A B C D

15. When an operator is checking the cutter clearance angle with a dial indicator on a .06-in.-wide land having a clearance of 4°, the indicator needle would move
    (A) .002 in.   (C) .006 in.
    (B) .004 in.   (D) .008 in.   15. A B C D

16. Which of the following is *not* used to check cutter clearance angles?
    (A) dial indicator   (C) Starrett clearance gage
    (B) Brown & Sharpe clearance gage (D) vernier protractor   16. A B C D

17. The proper land width is produced by grinding the
    (A) tooth face (C) secondary clearance
    (B) periphery   (D) primary clearance   17. A B C D

18. When sharpening a form-relieved cutter, grind it on the
    (A) periphery   (C) back of the tooth
    (B) tooth face (D) all of these   18. A B C D

19. When a plain helical milling cutter is being ground, the tooth rest is mounted on the
    (A) workhead   (C) upper table
    (B) saddle   (D) wheelhead   19. A B C D

20. Shell end mills should be ground on
    (A) a plain mandrel   (C) an expanding arbor
    (B) an expanding mandrel (D) the same arbor that is used for milling   20. A B C D

SCORE _____

20

TEST 68   Manufacture of Iron *(Unit 76)* and Steel

Most of the products that give us such a high standard of living, such as machines, automobiles, refrigerators, and stoves, are made mainly of ferrous metals—iron and steel. An understanding of these metals, their manufacture, and particularly their properties is useful to the machine shop worker.

Place the correct word(s) in the blank space(s) provided at the right-hand side of the page that will make the sentence complete and true.

1. __?__ ore, __?__ , and __?__ are loaded into a __?__ furnace to produce pig iron.

     1. _____

     _____

     _____

     _____

2. Since many open hearth furnaces and Bessemer converters are being phased out, they are being replaced by __?__ __?__ electric arc and __?__ __?__ furnaces.

     2. _____

     _____

     _____

     _____

3. The newest and a more efficient method of producing steel is by __?__ that produce steel faster and at lower cost.

     3. _____

4. In this process, the impurities are burned out of the steel by __?__ that is blown onto the molten metal through a water-cooled __?__ .

     4. _____

     _____

5. After the molten steel has been poured into the ladle, __?__ elements are added.

     5. _____

6. The electric furnace is used to produce line __?__ steel.

     6. _____

7. With the electric furnace, the __?__ , the amount of oxygen, and the atmospheric condition can be controlled.

     7. _____

8. Molten steel may be converted into __?__ or blooms by the continuous-casting process.

     8. _____

9. With this process, the molten metal is poured into a __?__ or reservoir at the top of the machine.

     9. _____

10. The molten metal then flows into the __?__ , which moves up and down to prevent hot metal from __?__ .

     10. _____

     _____

11. A solid skin is formed due to the __?__ action of the mold.

     11. _____

12. After the strand becomes solid throughout, it is cut to length by a __?__ cutting torch.

     12. _____

SCORE _____

20

TEST 69   Properties, Composition, and *(Units 76, 77, 78)* Identification of Metals

Although iron and carbon are the basic elements of steel, additions of alloying elements can change the structure of the steel and may also affect its machinability. The machinist should therefore know the effect of these elements on the steel and how it should perform in service, as well as the problems involved in machining the metal.

Place the correct word(s) in the blank space(s) provided at the right-hand side of the page that will make the sentence complete and true.

1. The amount of __?__ added to steel will determine its hardness and strength.

2. The addition of __?__ makes steel ductile and gives it good bending qualities.

3. Phosphorus is an undesirable element that makes steel __?__ .

4. Silicon is added to molten steel to remove the __?__ and oxides.

5. __?__ is an undesirable element that causes cracks in the steel when it is heated to a red color.

6. Sulfur or __?__ may be added to steel to give it __?__ cutting and good machining qualities.

7. Chromium in steel imparts __?__ and __?__ resistance.

8. Molybdenum permits cutting tools to retain their __?__ when hot.

9. Nickel improves the __?__ of steel and its resistance to fatigue, impact, and __?__ .

10. __?__ permits a cutting tool to maintain its cutting edge at red heat.

11. Carbon and chromium __?__ ductility of steel.

12. Medium-carbon steel contains from 0. __?__ percent to 0. __?__ percent carbon.

13. An alloy steel is one which contains __?__ or more metals.

14. High-strength, low-alloy steels can be machined, formed, or welded as easily as __?__ -carbon steels.

15. The SAE and __?__ numbers help to identify the type and composition of a steel.

16. Nonferrous metals contain no __?__ .

17. __?__ is that property of a metal that permits no permanent distortion before breaking.

18. __?__ is the ability of a metal to be permanently distorted or bent without breaking.

19. __?__ is the property of a metal to withstand shock or impact.

20. __?__ may be defined as the resistance of a metal to forcible penetration or bending.

21. __?__ is the ability of a metal to return to its original shape after any force acting on it has been removed.

1. _____

2. _____

3. _____

4. _____

5. _____

6. _____

  _____

7. _____

  _____

8. _____

9. _____

  _____

10. _____

11. _____

12. _____

  _____

13. _____

14. _____

15. _____

16. _____

17. _____

18. _____

19. _____

20. _____

21. _____

SCORE _____

25

TEST 70 Heat Treatment and Testing of Steel *(Units 77, 78)*

Any part, regardless of how accurately it is made, will fail in use if it is not made from the proper type of steel or correctly heat treated. The machinist must therefore have some knowledge as to the selection, heat treatment, and testing of steel.

Select the most appropriate answer for each question and circle the letter in the right-hand column that indicates your choice.

1. Reheating carbon steel to a desired temperature below its lower critical temperature and quenching in water or oil is known as
   (A) hardening
   (B) tempering
   (C) annealing
   (D) normalizing

   1.　A　B　C　D

2. Steel that contains just enough carbon to dissolve completely in the iron when the steel is heated to its critical range is called
   (A) eutectoid
   (B) hypoeutectoid
   (C) hypereutectoid
   (D) austenite

   2.　A　B　C　D

3. Alternate layers of ferrite and cementite form
   (A) austenite
   (B) martensite
   (C) alpha iron
   (D) pearlite

   3.　A　B　C　D

4. The point at which carbon steel changes into austenite is the
   (A) decalescence
   (B) recalescence
   (C) critical range
   (D) upper critical temperature

   4.　A　B　C　D

5. Heating steel to just above the upper critical temperature and cooling it in still air is called
   (A) annealing
   (B) carburizing
   (C) normalizing
   (D) tempering

   5.　A　B　C　D

6. Most water-hardening steels achieve a *maximum* hardness
   (A) for a depth of .12 in.
   (B) for a depth of .25 in.
   (C) for a depth of .50 in.
   (D) throughout

   6.　A　B　C　D

7. Steels with a high manganese content that are to be hardened should be quenched in
   (A) water
   (B) brine
   (C) oil
   (D) air

   7.　A　B　C　D

8. If an intricate part warps during quenching, it is best to
   (A) use a water-hardening steel
   (B) use an air-hardening steel
   (C) use an oil-hardening steel
   (D) quench it slowly in brine

   8.　A　B　C　D

9. High-speed steel toolbits can retain their cutting edge at red heat because of the addition of 18 percent
   (A) chromium
   (B) vanadium
   (C) manganese
   (D) tungsten

   9.　A　B　C　D

10. Air-hardening steels are often used on large workpieces because
    (A) it is more economical than using a liquid quench
    (B) of the faster cooling rate
    (C) large quenching tanks are not always available
    (D) full hardness can be achieved

    10.　A　B　C　D

11. Which of the following changes does *not* occur when 0.83 percent carbon steel is heated to about 1330°F?
    (A) The atoms rearrange themselves from face-centered cubes to body-centered cubes.
    (B) The steel becomes nonmagnetic.
    (C) A slight drop in the temperature of the workpiece is noticed.
    (D) The structure changes from pearlite to austenite.

    11.　A　B　C　D

12. Steel begins to harden when it is quenched after reaching the
    (A) upper critical temperature
    (B) lower critical temperature
    (C) recalescence point
    (D) carburizing temperature

    12.　A　B　C　D

continued on next page

13. To obtain *maximum* hardness with a hypereutectoid steel before quenching, heat it to

13.   A   B   C   D

(A) 50°F above the upper critical temperature
(B) 50°F above the lower critical temperature
(C) the recalescence point
(D) the decalescence point

14. When steel that has been properly heated is cooled rapidly,

14.   A   B   C   D

(A) it passes through the decalescence point
(B) it passes through the recalescence point
(C) pearlite changes to austenite
(D) austenite changes to martensite

15. If a steel having over 0.83 percent carbon content is heat-treated properly,

15.   A   B   C   D

(A) the steel will be harder than 0.83 percent steel
(B) the upper critical temperature will be raised
(C) there will be no change in hardness
(D) the wear resistance will be increased

16. Which of the following statements is *not true* of a Rockwell hardness tester?

16.   A   B   C   D

(A) It measures the amount of penetration of a diamond point.
(B) It is a direct-reading device.
(C) It requires a conversion table.
(D) It uses a 150-kg load.

17. The penetrator used with a Brinell hardness tester for ferrous metals is a

17.   A   B   C   D

(A) 1/16-in.-diameter hardened ball
(B) 10-mm-diameter hardened ball
(C) 110° conical diamond
(D) 1/8-in.-diameter hardened ball

18. The scleroscope hardness tester operates on the principle of the

18.   A   B   C   D

(A) rebounding of the diamond-tipped hammer
(B) depth of impression of the penetrator
(C) area of impression of the penetrator
(D) rebounding of the 10-mm hardened-steel ball

19. The tensile strength of a metal is the

19.   A   B   C   D

(A) amount it will stretch before breaking
(B) maximum amount of pull before breaking
(C) hardness and toughness
(D) area of the cross section divided by the load

20. When stress is applied to a specimen with a tensile tester, the needle will move uniformly with the load for a short time. When this uniform movement changes, the metal reaches its

20.   A   B   C   D

(A) ultimate strength
(B) yield point
(C) proportional limit
(D) elastic limit

SCORE _____

20

TEST 71 Cellular Manufacturing *(Unit 79)*

The basic concept of cellular manufacturing is in the integration of management practices with technological advances. To be truly successful requires thorough understanding of the causes and elimination of waste at all levels, and that means both operations and processes.

Place the correct word(s) in the blank space(s) provided at the right-hand side of the page that will make the sentence complete and true.

1. Cellular manufacturing helps to create a concept of __?__.

2. The layout of the equipment and the workstations is __?__ and __?__.

3. Five ways factories converting to cellular manufacturing benefit are __?__, __?__, __?__, __?__, and __?__.

4. Successful cellular manufacturing requires an understanding of __?__.

5. The term *process* is defined as __?__.

6. Production starts when __?__.

7. Two skills a team member should possess are __?__ and __?__.

8. Two machine tools commonly found in a manufacturing cell are __?__ and __?__.

9. Four main manufacturing systems used to provide productivity and flexibility are __?__, __?__, __?__, and __?__.

10. In cellular manufacturing, a group of similar parts is referred to as a __?__.

1. _____

2. _____
   _____

3. _____
   _____
   _____
   _____
   _____

4. _____

5. _____

6. _____

7. _____
   _____

8. _____
   _____

9. _____
   _____
   _____
   _____

10. _____

SCORE _____

20

TEST 72   Kaizen *(Unit 80)*

The journey to Lean is an ongoing one; it requires a strong commitment, a good organizational structure, and just plain hard work. In Lean manufacturing, this change for the better can result in gradual improvement of products, workplace efficiency, customer service, and reduction of waste.

Place the correct word(s) in the blank space(s) provided at the right-hand side of the page that will make the sentence complete and true.

1. The best way to describe the transformation of a factory to Lean manufacturing operation is __?__ .

2. The term *kaizen* is defined as __?__.

3. During the transition, management must provide __?__ and __?__.

4. The primary reason that a company introduces kaizen is __?__.

5. Six typical forms of waste are __?__, __?__, __?__, __?__, __?__, and __?__.

6. Two main causes of motion waste are __?__ and __?__.

7. Every process during manufacturing must include __?-?__ activity.

8. Four benefits of kaizen are __?__, __?__, __?__, and __?__.

9. It is important to hold kaizen events to __?__.

1. _____

2. _____

3. _____
   _____

4. _____

5. _____
   _____
   _____
   _____
   _____
   _____

6. _____

7. _____

8. _____
   _____
   _____
   _____

9. _____

SCORE _____

20

TEST 73   Kanban *(Unit 81)*

Pull/kanban is the part of Lean production that applies to the elimination of waste and all areas of a company can benefit from the application of Lean principles. Reduction of waste ensures lower costs, higher quality products, and better service and delivery.

Place the correct word(s) in the blank space(s) provided at the right-hand side of the page that will make the sentence complete and true.

1. The term *kanban* means __?__.

   1. _____

2. Five reasons for generating a Kanban card to __?__, __?__, __?__, __?__, and __?__.

   2. _____
   _____
   _____
   _____
   _____

3. The main purpose of the Lean system is __?__.

   3. _____

4. Three benefits of reducing waste in a company are __?__, __?__, and __?__.

   4. _____
   _____
   _____

5. The term *waste* as it applies to the Lean system of manufacturing means __?__.

   5. _____

6. The difference between pull and push systems is that in a pull system you __?__, whereas in a push system you __?__.

   6. _____
   _____

7. An example of the pull/kanban concept is __?__.

   7. _____

8. Six benefits of the pull/kanban system in a company are __?__, __?__, __?__, __?__, __?__, and __?__.

   8. _____
   _____
   _____
   _____
   _____
   _____

9. The 5S signals in workplace organization are __?__, __?__, __?__, __?__, and __?__.

   9. _____
   _____
   _____
   _____
   _____

   SCORE _____

   25

TEST 74 Total Productive Maintenance *(Unit 82)*

The main purpose of TPM is to ensure that all equipment required for production is operating at 100 percent efficiency at all times. Short daily inspections, cleaning, lubricating, and making minor adjustments can detect minor problems that can be corrected before they shut down a production line. TPM requires company support ranging from top executive to the shop floor personnel.

Place the correct word(s) in the blank space(s) provided at the right-hand side of the page that will make the sentence complete and true.

1. The term *total productive maintenance* can be defined as __?__.

2. Four types of maintenance are __?__, __?__, __?__, and __?__.

3. Two goals of the TPM program are to __?__ and __?__.

4. The three main objectives of the TPM program are to have __?__, __?__, and __?__.

5. The cost of machine breakdown is __?__ to __?__ % higher than maintenance cost.

6. Two types of machine breakdowns are __?__ and __?__.

7. Three causes of machine breakdowns are __?__, __?__, and __?__.

8. __?__ and __?__ are two types of machine deterioration.

9. __?__ tried to warn the American automotive industry that quality is important.

1. _____

2. _____
   _____
   _____
   _____

3. _____
   _____

4. _____
   _____
   _____

5. _____
   _____

6. _____
   _____

7. _____
   _____
   _____

8. _____
   _____

9. _____

SCORE _____
20

TEST 75   Value Stream Mapping *(Unit 83)*

Value stream mapping (VSM) is a method of recording a product's production path (materials and information) from door to door. It can serve as a starting point for management, engineers, production associates, schedulers, suppliers, and customers to recognize and identify waste and its causes.

Place the correct word(s) in the blank space(s) provided at the right-hand side of the page that make the sentence complete and true.

1. Value stream mapping is a ___?-?___ technique.

    1. _____
    _____

2. Production flow is the ___?___ of ___?___ through a factory.

    2. _____
    _____

3. Cross-training employees makes them ___?___ for other jobs.

    3. _____

4. The four stages of recording the activities in value-stream mapping are ___?___, ___?___, ___?___, and ___?___.

    4. _____
    _____
    _____
    _____

5. Three reasons why VSM is an important Lean organizational method are ___?___, ___?___, and ___?___.

    5. _____
    _____
    _____

6. To be effective, VSM relies on ___?___ and ___?___.

    6. _____
    _____

7. To improve and promote process learning, VSM links ___?___, ___?___, ___?___, and ___?___.

    7. _____
    _____
    _____
    _____

8. Two problems that may arise when implementing VSM are ___?___ and ___?___.

    8. _____
    _____

    SCORE _____

    20

TEST 76   Workplace Organization *(Unit 84)*

The workplace environment is important in helping to introduce Lean manufacturing. An untidy or disorganized workplace can lead to wasted energy, whether it be avoiding obstacles or searching for materials and tools. Many companies are working on the 5S system that consists of effective workplace organization and standard procedures that are favorable to Lean manufacturing principles.

Place the correct word(s) in the blank space(s) provided at the right-hand side of the page that will make the sentence complete and true.

1. The main problems of an untidy or disorganized workplace are __?__ and __?__ for materials.

     1. _____
     _____

2. Five things the 5S philosophy focuses on are __?__, __?__, __?__, __?__, and __?__.

     2. _____
     _____
     _____
     _____
     _____

3. The best way to turn ideas into success is to __?__ the __?__ of every employee.

     3. _____
     _____

4. __?__ and __?__ are two aspects that should be considered in the application of ergonomics.

     4. _____
     _____

5. Operating waste is __?__ __?__.

     5. _____

6. __?__, __?__, and __?__ are the three main things that 5S focuses on in manufacturing.

     6. _____
     _____
     _____

7. Four benefits of implementing the 5S system are __?__, __?__, __?__, and __?__.

     7. _____
     _____
     _____
     _____

SCORE _____

20

TEST 77 Computer Numerical Control *(Unit 86)*

Computer numerical control (CNC) has continually evolved as one of the major improvements in the manufacturing industry, resulting in the efficient manufacture of better products at lower prices. CNC enables any number of parts to be machined accurately and greatly reduces the possibility of human error.

Place the correct word(s) in the blank space(s) provided at the right-hand side of the page that will make the sentence complete and true.

1. The function of a computer used with CNC is to receive __?__ instruction in numerical form for processing, to operate the __?__ .

    1. _____
    _____

2. Four reasons for the wide acceptance of CNC throughout the world are the result of their __?__ , __?__ , __?__ , and __?__ .

    2. _____
    _____
    _____
    _____

3. The three primary axes used on a CNC machine are __?__ , __?__ , and __?__ .

    3. _____
    _____
    _____

4. When in the X axis of the coordinate system, a __?__ movement is always to the right of the zero or origin point.

    4. _____

5. Locate the five coordinate locations shown on the drawing.

    5. A _____ , _____
    B _____ , _____
    C _____ , _____
    D _____ , _____
    E _____ , _____

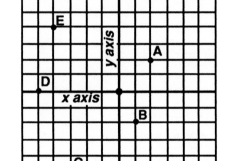

6. Computer numerical control programming falls into two distinct categories, __?__ and __?__ .

    6. _____
    _____

7. In the __?__ programming system, all dimensions or positions are given from a zero or reference point.

    7. _____

8. The reason that a closed loop is more reliable than an open loop system is because it is equipped with a __?__ system.

    8. _____

9. The two most common types of input media used for CNC machine tools are __?__ and __?__ .

    9. _____
    _____

10. G-codes are called __?__ commands.

    10. _____

11. The most commonly used forms of interpolation are __?__ and __?__ .

    11. _____
    _____

12. The __?__ code is used for rapid movement.

    12. _____

    SCORE _____

25

TEST 78 (Part 1) Turning or *(Unit 87)* Chucking Centers

Turning or chucking centers can produce accurate work automatically, at high production rates. These machines are used to produce parts similar to those made on lathes.

Place the correct word(s) in the blank space(s) provided at the right-hand side of the page that will make the sentence complete and true.

1. __?__ centers are designed to machine workpieces normally held in some form of chuck.

2. These centers fall into three categories, namely, the __?__ , __?__ , and the __?__ / __?__ center.

3. Some machines have an upper and lower turret that can be used to machine __?__ and __?__ diameters simultaneously.

4. The rugged, heavy one-piece casting is slanted to remove __?__ easily and for convenience in __?__ or __?__ the workpiece.

5. Three main framework components of a turning center are the __?__ , __?__ , and __?__ .

6. The main CNC components of a turning center are the __?__ and the __?__ .

7. Tools that rotate while in the tool turret are called __?__ tools.

8. A turning center spindle that reacts like an indexing head is called a __?__ spindle.

9. On turning centers, some controls program the X axis in __?__ dimensions, while others use __?__ dimensions.

10. The programmer may establish the part Z0 at the __?__ or __?__ end of the part.

1. _____

2. _____
   _____
   _____
   _____

3. _____
   _____

4. _____
   _____
   _____

5. _____
   _____
   _____

6. _____
   _____

7. _____

8. _____

9. _____
   _____

10. _____
    _____

SCORE _____

20

TEST 78 (*Cont'd*) (Part 2) Turning or *(Unit 87)* Chucking Centers

This project consists of machining two parallel diameters and chamfers with the workpiece held in a chuck. This provides practice in linear interpolation where the program is along the X axis (tool infeed) and the Z axis (longitudinal feed). It involves the use of rapid feed, setting depths of cuts, and tools retract.

Program Notes

1. MATERIAL: brass 2.0 in. diameter (CS500)

2. CUTTING TOOL: diamond-shaped carbide

3. Program in the absolute mode

4. All programming begins at the zero or reference point (XZ) at the centerline and the right-hand face of the part

5. All tool changes are at X 1.25 Z.25

6. Use diameter programming

7. Feed rate .010 in/rev

8. Use G82 fixed cycle for spot drill

9. Use G81 fixed cycle for drill

PROGRAM SEQUENCE

%	Rewind/parity check
N010	G20 T0100
N020	
N030	
N040	
N050	
N060	
N070	
N080	
N090	
N100	
N110	
N120	
N130	
N140	
N150	
N160	
N170	
N180	
N190	
N200	
N210	
N220	
N230	
%	rewind/stop

SCORE   25

TEST 79 (Part 1) Machining Centers *(Unit 88)*

CNC machining centers can produce accurate work automatically at high production rates. Machining centers produce parts normally made on horizontal and vertical milling machines.

Place the correct word(s) in the blank space(s) provided at the right-hand side of the page that will make the sentence complete and true.

1. Three main types of machining centers are __?__ , __?__ , and __?__ types.

    1. _____
    _____
    _____

2. The __?__ -column machining center is equipped with one or two tables, and the column (and the cutter) move while the work is being machined.

    2. _____

3. The __?__ -column machining center is generally equipped with a __?__ shuttle where a workpiece can be mounted while another is being machined.

    3. _____
    _____

4. Five framework components of the machining center are the __?__ , __?__ , __?__ , __?__ , and __?__ .

    4. _____
    _____
    _____
    _____
    _____

5. Each tool held in an automatic tool changer is identified by either the __?__ number or the storage __?__ number.

    5. _____
    _____

6. The advantages of torque controlled machining are that as the cutting tool starts to dull, the __?__ will be decreased or the operation will be __?__ .

    6. _____
    _____

7. When a tool changer rotates in the shortest direction to index a tool, it is known as __?__ .

    7. _____

8. The most common tool holder used on machining centers has a __?__ and self-releasing taper shank.

    8. _____

9. In order to save valuable setup and production time, and to increase the flexibility of a precision vise, a __?__ - __?__ system can be added.

    9. _____ - _____

10. Two of the safest ways to check a program for correctness are __?__ and __?__ run.

    10. _____
    _____

SCORE _____

20

TEST 79 (*Cont'd*)   (Part 2) Machining Centers *(Unit 88)*

This project involves the milling (programming the cutter path) of the part boundary, spot drilling, and drilling the holes, using the G81 fixed cycle for drilling holes in sequence. Proper speeds and feeds should be calculated and programmed for the type of material being cut.

PROGRAM NOTES

1. MATERIAL: Aluminum 1/4 in. thick (CS 500)

2. High-speed steel cutting tools are being used
   (a) 1/2 in. diameter four flute-end mill
   (b) 1/2 in. diameter spot drill
   (c) 1/2 in. diameter drill.

3. Program clockwise in the absolute mode

4. All programming begins and ends at the machine zero

5. Use coolant when machining

6. Part zero indicated on drawing

PROGRAM SEQUENCE

%	Rewind/parity check
N010	G20 G90 (inch) (absolute)
N020	G17 G40 G80
N030	T01 M06 (tool change)
N040	_____
N050	_____
N060	_____
N070	_____
N080	_____
N090	_____
N100	_____
N110	_____
N120	_____
N130	_____
N140	_____
N150	_____
N160	_____
N170	_____
N180	_____
N190	_____
N200	_____
N210	_____
N220	_____
N230	_____
N240	_____
N250	_____
%	(rewind/stop code)

SCORE   25

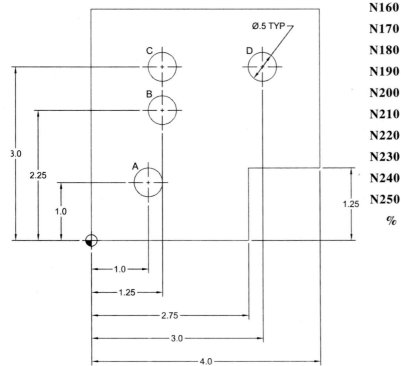

TEST 80   CAD/CAM *(Unit 89)*

CAD/CAM is the marriage of the computerized forms of drafting/design and manufacturing. In the last ten years probably no other phase of manufacturing has progressed as quickly as PC-based CAD/CAM technology.

Place the correct word(s) in the blank space(s) provided at the right-hand side of the page that will make the sentence complete and true.

1. ___?___ , a rapidly spreading technology, links the programming abilities of the   1. _____
    modern computer to material processing CNC machines, lathes, mills, grinders, punches, etc.

2. ___?___ is the entire process of the art to part concept. It covers the creation of a part   2. _____
    from the initial design to the finished machine part produced on a numerically controlled
    machine tool.

3. The first step of the CAD/CAM process is the ___?___ of the geometry.   3. _____

4. ___?___ geometry, the traditional type of geometry, has as its purpose the communication   4. _____
    of design intent.

5. ___?___ geometry is created from a CAD or drafting point of view with every single detail and   5. _____
    edge of the part represented.

6. In the viewing and verifying step, once the parameters have been set, the toolpath is displayed   6. _____
    on the screen in a process called ___?___ .

7. When all of the toolpaths have been assigned and visually verified, the next step is to turn   7. _____
    the toolpath files into G & M codes by a process called ___?___ .

8. Raised material defined by an enclosed contour inside a pocket may be left standing in the   8. _____
    cavity, if desired. The materials left standing are commonly referred to as ___?___ .

9. The purpose of a(n) ___?___ toolpath is to create holes, and/or enlarge them.   9. _____

10. The purpose of a(n) ___?___ toolpath is to remove the material from a part that is destroyed   10. _____
    as a surface or solid.

                                                        SCORE   _____
                                                                      10

TEST 81   Advanced Digital Manufacturing *(Unit 90)*

In the highly competitive manufacturing industry, companies need to build products better and cheaper. Time to market is critical and reducing that time through the use of Advanced Digital Manufacturing can increase a company's share of the market.

Place the correct word(s) in the blank space(s) provided at the right-hand side of the page that will make the sentence complete and true.

1. CAD is short for __?__ .

   1. _____

2. CAM is short for __?__ .

   2. _____

3. __?__ is a reliable, cost-effective method of making end-use parts for pre-production or production applications.

   3. _____

4. __?__ and __?__ are the two methods of Advanced Digital Manufacturing.

   4. _____

   _____

5. The major component of an ADM system that produces highly detailed three-dimensional parts with fine surface quality is the __?__ system.

   5. _____

6. The major component of an ADM system that is primarily used to produce functional parts for use in pre-production and production applications is the __?__ system.

   6. _____

7. __?__ is the component of an ADM system that uses hot melt, ink jet technology to build 3-D models in successive layers using thermoplastic materials.

   7. _____

8. In 2000, __?__ developed the prototype of the Advanced Digital Manufacturing system, which evolved into multiple solid-modeling imaging tools, new material systems, and solid-imaging technologies.

   8. _____

9. The two general classes of materials used in solid-imaging systems are __?__ and __?__ .

   9. _____

   _____

10. __?__ , the next logical step in the manufacturing process after CAD/CAM, was created for the design and production of prototype models in order to reduce or eliminate manufacturing errors and bring products to the market faster and at lower costs.

    10. _____

SCORE _____

12

TEST 82   Cryogenic Treatment/Tempering *(Unit 91)*

Cryogenic treatment/tempering is a one-time, permanent process that improves the physical and mechanical properties of various materials. The process can extend the lives of products such as drills, taps, reamers, broaches, mills, dies, and others.

Place the correct word(s) in the blank space(s) provided at the right-hand side of the page that will make the sentence complete and true.

1. __?__ is the process of cooling parts to approximately −110 degrees Fahrenheit by placing them in vats of alcohol cooled by dry ice.

2. __?__ is the process of cooling parts to −300 degrees Fahrenheit to improve the properties of materials.

3. __?__ are the maximum stresses to which materials may be subjected without any permanent strain remaining upon complete release of strength.

4. After quenching from high temperature, austenite becomes __?__ , a different crystalline form of steel.

5. A typical cycle of the cryogenic process lasts about __?__ days.

6. __?__ in steel comes from the cooling of uneven sections and machining—creating complex, invisible random patterns.

7. The __?__ modulus is the slope of the stress-strain curve at a specified point.

8. __?__ takes place when the entire mass of a material is at an equal temperature (core and surface) and cycled through a wide temperature range.

9. The __?__ cryogenic process physically transforms the microstructure into a new, more refined, uniform substructure without exposing the material to liquid nitrogen.

10. __?__ is a measure of the rigidity of metal—the ratio of stress, within proportional limits, to corresponding strain.

1. _____

2. _____

3. _____

4. _____

5. _____

6. _____

7. _____

8. _____

9. _____

10. _____

SCORE _____

10

TEST 83   QQC Diamond Coating *(Unit 92)*

The QQC process can deposit a uniform layer of diamond on almost any type of material. This laser process can be completed quickly and produces pure diamonds.

Place the correct word(s) in the blank space(s) provided at the right-hand side of the page that will make the sentence complete and true.

1. __?__ is the process that uses carbon dioxide from the air as the carbon source and subjects it to a combination of lasers.

   1. _____

2. A high degree of hardness, thermal conductivity, electronic mobility, and sound velocity are some of the qualities of __?__ .

   2. _____

3. In what decade were man-made diamonds first used in manufacturing for machining and grinding difficult-to-cut materials?

   3. _____

4. The synthesis process, invented by the General Electric Company, which requires pressures of 1 million pounds per square inch and temperatures of about 2000 degrees Fahrenheit, is the __?__ process.

   4. _____

5. The first step in the __?__ process is to produce atomic hydrogen from the diatomic hydrogen molecule.

   5. _____

6. The primary market focus for the developers of the CVD process has been __?__ .

   6. _____

7. In the QQC process, laser energy is provided by a combination of these three lasers: 1. __?__ 2. __?__ 3. __?__ .

   7. _____

   _____

   _____

8. The thickest layer of diamond made by the QQC process has been __?__ microns.

   8. _____

9. The preferred cutting tool material for machining cast iron is __?__ .

   9. _____

10. Metallurgist __?__ is credited with developing the QQC diamond coating process.

   10. _____

   SCORE _____

   12

TEST 84   Direct Metal Deposition *(Unit 93)*

Direct metal deposition (DMD), a rapid prototyping process, is a form of rapid tooling that makes parts and molds from metal powder. It melts the powder into a computer-aided design of the part and is then solidified in place. The DMD process improves the properties of materials in less time and at lower costs than traditional fabrication technologies.

Place the correct word(s) in the blank space(s) provided at the right-hand side of the page that will make the sentence complete and true.

1. __?__ is the rapid prototyping process that makes parts and molds from metal powder that is melted by a laser to a CAD design of the part and then solidified in place.

    1. _____

2. DMD closely resembles conventional rapid prototyping processes but differs in that __?__ and __?__ can be melted rather than plastic polymers.

    2. _____
       _____

3. DMD is the blending of these five technologies: 1. __?__ 2. __?__ 3. __?__ 4. __?__ 5. __?__ .

    3. _____
       _____
       _____
       _____
       _____

4. A typical DMD job starts at the secure website of __?__ , where customers post CAD files.

    4. _____

5. The Big Three of Manufacturing in the DMD process are: 1. __?__ 2. __?__ and 3. __?__

    5. _____
       _____
       _____

6. Using the __?__ process, manufacturing companies can produce rapid metal prototypes instead of plastic SLA models.

    6. _____

7. In the five steps of the molding process, the most time consuming is __?__.

    7. _____

8. In the DirecTool process, the first stage is __?__ .

    8. _____

9. Due to the strong metallurgical bond and fine uniform microstructures it produces, the DMD process is ideally suited for repair work in the __?__ industry.

    9. _____

10. Using the DirecTool process, no matter how complex the CAD file, it can be turned into production tooling in no more than __?__ hours.

    10. _____

             SCORE _____
                        17

TEST 85   e-Manufacturing *(Unit 94)*

Manufacturers must improve their productivity in order to remain competitive. To accomplish this they must take advantage of the automation technology now available. e-Manufacturing is about improving the communications to and from the factory floor with the customer.

Place the correct word(s) in the blank space(s) provided at the right-hand side of the page that will make the sentence complete and true.

1. Connecting machine tools on the shop floor into an overall __?__ will release (unleash) the information from each machine and allow management to increase profits.

 1. _____

2. __?__ creates a secure, open architecture platform that turns every machine tool into a node on the corporate network.

 2. _____

3. A machine tool as a node on the corporate network becomes a(n) __?__ that connects the point of production to management's information system—the supply and demand chain—in real time.

 3. _____

4. By introducing the concept of an ever-optimizing __?__ onto the shop floor, e-Manufacturing concepts will bring a new class of service to the entire corporation.

 4. _____

5. The value of extending Ethernet connectivity to the factory floor will result in these three benefits: __?__ , __?__ , and __?__ .

 5. _____

 _____

 _____

Decide whether each statement is true or false and circle the letter in the right-hand column that indicates your choice.

6. The main reason why a Web-enabled factory floor is so important is that it allows for the theory of constraints.

 6. T   F

7. Platinum Maintenance Service provides end users with remote diagnostics.

 7. T   F

8. The technology of networking tools over the Internet began in the 21st century.

 8. T   F

9. CNCs that work with a proprietary (closed) architecture platform are able to communicate with other CNCs.

 9. T   F

10. Although extending Ethernet connectivity to the factory floor will result in many benefits, costs will increase.

 10. T   F

SCORE   _____

12

TEST 86   STEP-NC and Internet Manufacturing *(Unit 95)*

Under development since 1984, STEP-NC has the potential for changing the way products are manufactured throughout the world. The standard, used to define data for numerical control machine tools, has the support of major companies like General Electric, Boeing, General Motors, and others.

Place the correct word(s) in the blank space(s) provided at the right-hand side of the page that will make the sentence complete and true.

1. __?__ is an extensible, comprehensive, international data standard for product data created by an international team of data experts.

    1. _____

2. STEP-NC replaces __?__ , the programming language that numerically controlled machines were using for nearly 50 years.

    2. _____

3. In the STEP-NC framework, which ISO model is used for milling machines? __?__

    3. _____

4. In the STEP-NC framework, which ISO model is used for EDM machine processing? __?__

    4. _____

5. In the STEP-NC framework, which ISO model is used for turning machines? __?__

    5. _____

6. STEP-NC changes the way manufacturing is done by defining data as __?__ , that is, a library of specific operations that might be performed on a CNC machine tool.

    6. _____

7. __?__ , a forerunner to STEP-NC, first appeared about 25 years ago when designers and engineers began to use computers to create designs.

    7. _____

8. Under the auspices of the ISO, the __?__ was established. Its goal was to define the methods for creating product data models that could be understood by computers.

    8. _____

9. STEP uses its own language called __?__ .

    9. _____

10. The __?__ project supports a three-stage, process-functional design and delivers data produced by the process to an intelligent controller.

    10. _____

SCORE _____

10

TEST 87   Optical Laser Vision Measurement *(Unit 96)*

For nearly 100 years manufacturers have been using various types of optical measuring instruments. Now video microscopes, laser measuring and tool setting instruments, optical comparators, and coordinate measuring machines are commonly used.

Place the correct word(s) in the blank space(s) provided at the right-hand side of the page that will make the sentence complete and true.

1. The most commonly used probe is the __?__ .   1. _____

2. In two-dimensional probing head systems, the __?__ mode is used for profiling.   2. _____

3. In two-dimensional probing head systems, the x and y axes are free to move in the __?__ mode.   3. _____

4. The __?__ probing head is used for measuring and inspecting convoluted (complex) part surfaces such as gears and cams.   4. _____

5. The __?__ , or shadowgraph, is a fast, accurate means of measuring or comparing the shape or accuracy of a workpiece with a master.   5. _____

6. The __?__ , a very precise and reliable tool for performing high-speed measurements on stationary or moving targets on production lines, uses a high-intensity LED light source and an HL-CCD.   6. _____

7. The __?__ , a highly accurate method of measurement, has its workpiece located in the center of the laser beam, creating a shadow in the path of the scanning beam, which when detected enables the unit to determine the edge of the part.   7. _____

8. __?__ are becoming more popular since they can be used outside of a clean environment-controlled room, allowing them to be used on the shop floor where measurements and corrections can be made while the workpiece is still in the machine.   8. _____

9. The __?__ system is a powerful measuring tool providing the user with online documentation and archiving, repeatability, and image processing and manipulation.   9. _____

10. __?__ provide a means of electronically enhanced optical-image magnification for small parts with three-dimensional characteristics, such as components mounted on a circuit board.   10. _____

SCORE _____

10

TEST 88 Electrical Discharge Machining *(Unit 97)*

Electrical discharge machining has been used for a long time. EDM is the process used to remove metal through the action of an electrical discharge of short duration and high current density between the tool or wire and the workpiece.

Place the correct word(s) in the blank space(s) provided at the right-hand side of the page that will make the sentence complete and true.

1. EDM removes the metal by means of a(n) __?__ .

2. In EDM, the electrode is made to the shape of the __?__ required and must be made of electrically __?__ material.

3. EDM uses a(n) __?__ current of __?__ voltage and __?__ amperage.

4. In EDM, both the workpiece and the tool are immersed in a(n) __?__ fluid.

5. In EDM, the cutting tool (electrode) is made from electrically conductive material, usually __?__ .

6. In EDM, a(n) __?__ maintains a gap of about .0005 to .001 inch between the electrode and the work, preventing them from coming into contact with each other.

7. The two most common types of electrical discharge power supplies are __?__ and __?__ .

8. Which type of electrical discharge power supply is used almost exclusively by American manufacturers? __?__

9. __?__ is the amount by which the cavity in the workpiece is cut larger than the size of the electrode used in the machining process.

10. The four main operating systems, or components, of wire-cut electrical discharge machines are the servomechanism, the dielectric fluid, the machine control unit, and the __?__ .

1. _____

2. _____

_____

3. _____

_____

_____

4. _____

5. _____

6. _____

7. _____

_____

8. _____

9. _____

10. _____

SCORE _____

14

TEST 89　　Robotics　*(Unit 98)*

Since first being used in the 1960s, industrial robots have become more sophisticated and are used in many more applications. Among these other applications, industrial robots are used in all types of manufacturing and assembly.

Place the correct word(s) in the blank space(s) provided at the right-hand side of the page that will make the sentence complete and true.

1. A(n) __?__ is a programmable, multifunctional manipulator designed to move material parts, tools, and devices through various programmed motions for the performance of a variety of tasks.

2. Of the four manufacturing areas where robots find wide use, __?__ is the one where robots are used to carry out operations such as metalizing, seam welding, spot welding, and other operations where the robot can use tools to carry out a manufacturing operation.

3. The simplest type of robot has a single arm equipped with a set of __?__ that are used to perform an operation, or to pick up and move something.

4. The part of the robot that uses tools and moves parts is called the __?__ .

5. The part of the robot that provides electricity for its motor(s) is called the __?__ .

6. The part of the robot that controls the actions of the robot is called the __?__ .

7. In which of the four modes of programming robots is the robot taught by leading the arm through the necessary movements by an operator? __?__

8. In which of the four modes of programming robots is the robot movement controlled by the computer program? __?__

9. Since the 1960s some of the most common applications for industrial robots have been for what has been described as the "Three Ds." These are __?__ , __?__ , and __?__ .

10. Besides the "Three Ds," industrial robots have been commonly used for applications called the "Three Hs." These are __?__ , __?__ , and __?__ .

1. _____

2. _____

3. _____

4. _____

5. _____

6. _____

7. _____

8. _____

9. _____

_____

_____

10. _____

_____

_____

SCORE _____

14

TEST 90   Manufacturing Intelligence   *(Unit 99)*

Today's global economy requires manufacturers to produce parts faster, more accurately, and at lower manufacturing costs. As a means of meeting these goals, manufacturers are turning to open-network-based tools and new manufacturing technologies. They hope to use real-time manufacturing intelligence to achieve optimal performance with a minimum of downtime.

Place the correct word(s) in the blank space(s) provided at the right-hand side of the page that will make the sentence complete and true.

1. __?__ refers to having access to production events over a network the moment they occur.   1. _____

2. __?__ is a Web-based solution that bridges the gap between traditional manufacturing execution systems (MES) and the systems that manage business and supply activities across multiple facilities.   2. _____

3. The __?__ , created by the Mazak Corporation, provides a comprehensive information-gathering system that starts with the design of a product, continues through every stage of the manufacturing process, and includes customer service.   3. _____

4. In the Cyber Management System, the information assembled through the __?__ is available to everyone involved in the manufacturing process.   4. _____

5. In the Cyber Management System, __?__ is the software component used because it can easily create machining programs on Windows-based personal computers.   5. _____

6. In the Cyber Management System, the __?__ is used for the online collection of shop floor data and production schedule monitoring.   6. _____

7. The component of the Cyber Management system that provides for online tool status monitoring, tool life management, and the required tooling list is __?__ .   7. _____

8. The component of the Cyber Management System that shows the real-time operation status of the networked machines, whether they are in the operation, alarm, or idle mode, is the __?__ .   8. _____

9. __?__ , created by Executive Manufacturing Technologies, contains a complete suite of advanced analytical tools that deliver complex data from across the plant floor to decision makers who can then make timely and accurate decisions.   9. _____

10. __?__ software, created by POLYPLAN Technologies, is a manufacturing process management (MPM) desktop solution that captures and promotes the reuse of manufacturing know-how to everyone involved in the development and execution of the manufacturing process.   10. _____

SCORE _____

10

TEST 91   Multitasking Machines *(Unit 100)*

Multitasking machines are being used more and more by manufacturers in order to provide better products faster and cheaper. A wide variety of multitasking machines are available including single-spindle machines, single vertical turret machines, twin-turning spindle machines, as well as others.

Place the correct word(s) in the blank space(s) provided at the right-hand side of the page that will make the sentence complete and true.

1. The __?__ network allows management to operate a factory in real time, providing accessibility to machine data, machining programs, fixture data, tool data, production schedules, and other data.

    1. _____

2. The __?__ provides a comprehensive information-gathering system that starts with the design of a product, continues through every stage of the manufacturing process, and includes customer service.

    2. _____

3. __?__ is defined as the ability to perform various manufacturing operations without manual intervention.

    3. _____

4. The information gathered by the Cyber Management System is assembled through the __?__ , which is available to everyone involved in the manufacturing process.

    4. _____

5. Fewer machines are required, time is saved, more accurate parts are made, and cash flow is improved are some of the advantages to a manufacturer using a(n) __?__ machine.

    5. _____

6. When using the Mazak Integrex Model 200-III for construction, the __?__ , 64-bit RISC processor provides complete shop floor communication and efficiency in real time.

    6. _____

7. Three accessories and optional equipment available to make multitasking machines a true combination of information technology (IT) and manufacturing technology (MT) are __?__ , __?__ , and __?__ .

    7. _____
       _____
       _____

8. The Mazak __?__ is a turning center that can perform secondary machine operations as well as being a powerful multitasking machine with great flexibility.

    8. _____

9. In tool data management, tool data is tied to a(n) __?__ mounted in each tool's retention stud. The tool data is automatically downloaded over the network when the device is read.

    9. _____

10. The __?__ , a cutting tool that can perform 12 machining operations without a tool change, is designed to complement multitasking Integrex machine tools.

    10. _____

                SCORE _____

                                 12

SECTION 2

PROJECTS

MACHINING PROCEDURES FOR ROUND WORKPIECES

Most of the work produced in a machine shop is round and is turned to size on a lathe. In industry, much of the round work is held in a chuck, while a larger percentage of work in training programs is machined between centers because of the need to reset work more often. In either case, it is important to follow the correct machining sequence of operations to prevent spoiling work, which often happens when incorrect procedures are followed. All dimensions are shown in decimal inches.

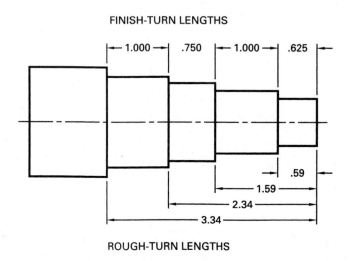

FINISH-TURN LENGTHS

ROUGH-TURN LENGTHS

FIG. C

GENERAL RULES FOR ROUND WORKPIECES

1. Rough-turn all diameters to within .03 in. (0.80 mm) of the size required.
   - Machine the largest diameter first and progress to the smallest.
   - If the small diameters are rough-turned first, it is quite possible that the work may bend when the large diameters are machined.

2. Rough-turn all steps and shoulders to within .03 in. (0.80 mm) of the length required (Fig. C).
   - Be sure to measure all lengths from one end of the workpiece.
   - If all measurements are not taken from the end of the workpiece, the length of each step will be .03 in. (0.80 mm) shorter than required.

3. If special operations such as knurling or grooving are required, they should be done next.

4. Cool the workpiece before starting the finishing operations.
   - Metal expands from the friction caused by the machining process, and any measurement taken while work is hot will be incorrect.
   - When the workpiece is too cold, the diameters of round work will be smaller than required.

5. Finish-turn all diameters and lengths.
   - Finish the largest diameter first, and then finish the shoulder to length.
   - Repeat this procedure, starting with the next largest diameter and working down to the smallest diameter.

There are many variations in size and shape of flat workpieces; therefore, it is difficult to give specific machining rules to follow that would apply to each. Some general rules are listed; however, they may have to be changed to suit particular workpieces. All dimensions are shown in decimal inches.

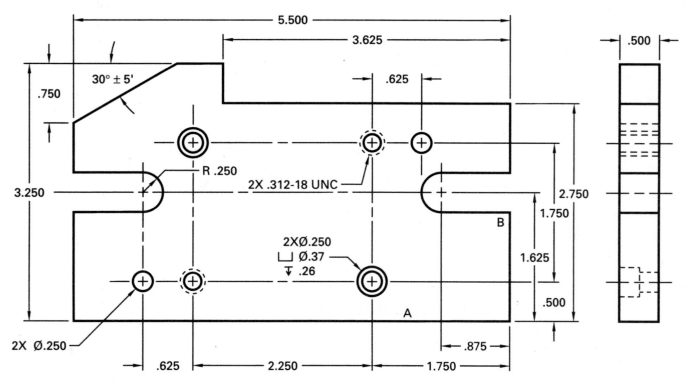

FIG. D

1. Select and cut off the material a little larger than the size required.

2. Machine all surfaces to size in a milling machine.

3. Lay out the physical contours of the part, such as angles, steps, radii, etc.

4. Lightly prick-punch the layout lines that indicate the surfaces to be cut.

5. Remove large sections of the workpiece on a contour bandsaw.

6. Machine all forms such as steps, angles, radii, and grooves.

7. Lay out all hole locations and, with dividers, scribe reference circles for each hole.

8. Drill all holes and tap those that require threads.

9. Ream holes.

10. Surface-grind any surface that requires it.

OPERATIONS SEQUENCE FOR A SAMPLE FLAT PART

The part shown in Fig. D is used only as an example to set forth a sequence of operations that should be followed when machining similar parts. These are not meant to be hard-and-fast rules but only act as a guide for the student.

The sequence of operations suggested for the sample part shown in Fig. D may be different from those suggested for machining a block square and parallel because:

1. The part is relatively thin and has a large surface area.

2. Since at least .12 in. (3 mm) of work should be above the vise jaws, it is difficult to use a round bar between the work and movable jaw for machining the large flat surfaces.

3. A small inaccuracy (out-of-squareness) on the narrow edge will create a greater error when the large surface has been machined.

Procedure

1. Cut off a piece of steel .62 × 3.38 × 5.62-in. (16 × 86 × 143-mm) long.

2. In a milling machine, finish one of the larger surfaces (face) first.

3. Turn the workpiece over and machine the other face to .50-in. (13-mm) thick.

4. Machine one edge square with the face.

5. Machine an adjacent edge square (90°) with the first edge.

6. Place the longest finished edge (A) down in the machine vise and cut the opposite edge to 3.25-in. (83-mm) wide. Leave .010 in. (0.25 mm) on each surface to be ground.

7. Place the narrower finished edge (B) down in the machine vise and cut the opposite edge to 5.50-in. (140-mm) long.

8. With edge A as a reference surface, lay out all the horizontal dimensions with an adjustable square, surface gage, or height gage.

9. With edge B as a reference surface, lay out all the vertical dimensions with an adjustable square, surface gage, or height gage.

10. Use a bevel protractor to lay out the 30° angle on the upper left-hand edge.

11. With a divider set to .25 in. (6 mm), draw the arcs for the two center slots.

12. With a sharp prick punch, lightly mark all the surfaces to be cut and the centers of all hole locations.

13. Center-punch, drill, and ream .50-in. (13-mm) diameter holes for the two center slots.

14. On a vertical bandsaw, cut the 30° angle to within .03 in. (0.8 mm) of the layout line.

15. Place the workpiece in a vertical mill and machine the two .50-in (13-mm) slots.

16. Machine the step on the top edge of the workpiece.

17. Set the work to 30° in the machine vise and finish the 30° angle.

18. Prick-punch the hole locations, scribe reference circles, and then center-punch all hole centers.

19. Center-drill all hole locations.

20. Drill and counterbore the two .250-in. holes.

21. Tap drill the .312-in. 18 UNC thread holes (F drill).

22. Drill the .250-in. (6-mm) ream holes to .23 in. (5.5 mm).

23. Countersink all holes to be tapped slightly larger than their finished size.

24. Ream the .250-in. (6-mm) holes to size.

25. Tap the .312-in. 18 UNC holes.

The drill size part of this gage is useful for checking some of the more common drill sizes to be sure that errors are not made by using the incorrect drill. The drill angle part of the gage can be used to check the angles and lengths of each cutting edge so that the drill will produce holes at the right locations and as close to size as possible.

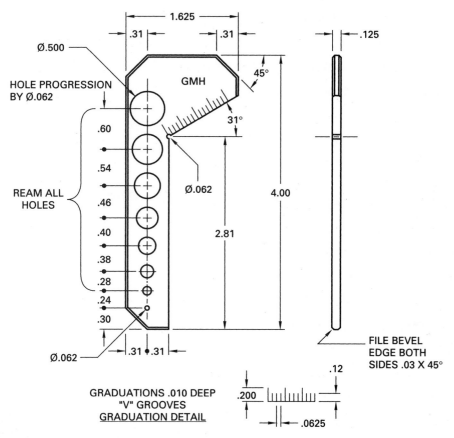

FIG. 1-1

OBJECTIVES

This project should provide the following learning experiences:

1. Laying out part outlines and hole locations with a combination set.

2. Drilling and reaming various size holes to accurate locations.

3. Using the file for short surfaces and beveled edges.

4. Learning how to graduate tools and instruments.

Operations

1. Cut off one piece of .125 × 1.75 × 4.12-in. long cold rolled steel (C.R.S.).

2. Place the piece in a vertical milling machine vise and cut only on one side to machine the width to 1.625 in.

3. Set the part, along with a piece of .500-in. thick steel about 3.750-in. long, vertically in the vise.

4. Use a square to set the part vertically in the vise and clamp it together with the plate in the vise.

5. Use a C-clamp to hold the part and the steel plate together.
(This will prevent the .125-in.-thick part from springing during machining.)

6. Machine the end smooth and square.

7. Remove the burrs and reset the work in the vise using the same procedures as in steps 4 and 5.

8. Machine the part to 4.000-in. length.

9. Apply layout dye to one of the flat surfaces.

10. Lay out the outline of the drill gage using a combination square and a sharp scriber.

11. Lightly prick-punch the outline of the part.

12. Lay out the locations of the eight holes.

13. Lightly prick-punch, check the locations for accuracy, correct if necessary, and then center-punch all locations.

14. Lay out the location of the .062-in. hole on the right side of the gage.

15. Drill the .062-in. hole through the part.

16. Use a center drill to spot the locations of the eight holes.

17. Drill the .062-in. hole through the part.

CAUTION: For the .062-in. hole, use a center drill with a drill smaller than .062-in., or lightly spot this hole.

18. Use a drill .015 in. smaller than each hole to drill the holes through the part.

19. Countersink each hole about .04-in. wide on each side of the part.

20. Ream all holes to size except the .062-in. hole.

21. On a contour bandsaw, cut the outline of the gage (corners, inside edge, and angle), staying away .03 in. from the layout line.

22. Set the gage in the machine vise at 45° for each corner and cut to the layout lines.
   • Have the layout line within .125 in. of the vise to prevent bending the part during machining.
   or
   • Set the part in a bench vise so that the layout line is aligned exactly with the top of the vise jaw and file each surface to size.

23. Set the gage in the vise on its left edge and machine the .625-in. width to size.

24. Set the gage in the vise and cut the 31° angular surface to the layout line.

25. Lay out and file the .03-in. beveled edges as indicated on the drawing.

26. Clamp the gage on a vertical mill table or in a holding fixture in preparation for graduating the scale.

27. Align the angular surface of the gage parallel to the table travel.

28. Insert a sharp-pointed 60° hardened tool in the machine spindle.

29. Locate the center point of the tool on the angular edge and at the center of the .062-in. drilled hole.

30. Set the table and crossfeed collars to zero.

31. Use a paper feeler to set the point of the tool to the top surface of the gage.

32. Raise the table .010 in. for the depth of the graduations.

33. Use the graduations on the table and crossfeed to scribe the lines to length and width.

34. Remove all burrs and polish all surfaces to finish the gage.

TACK HAMMER - PROJECT #2

Operations for Machining the Handle

1. Cut off one piece of hot-rolled machine steel .75-in. diameter × 8.80-in. long.

2. Grind or file a flat .25-in. wide × .75-in. long, as shown in Fig. 2-1.

3. Stamp your initials on the ground spot.

4. Lay out the center holes on each end.

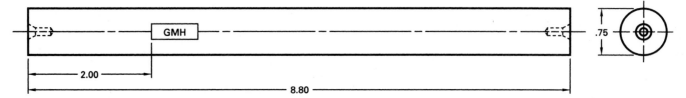

FIG. 2-1

5. Check center hole layout.

6. Lay out finished lengths 8.75 in. on the work.

7. Drill the center holes in each end of the work.

8. Mount work between lathe centers.

9. Face work to 8.75-in. length.

10. Re-center the ends, if necessary.

11. Lay out length of shoulder (4.75 in.) from the end opposite the initials.

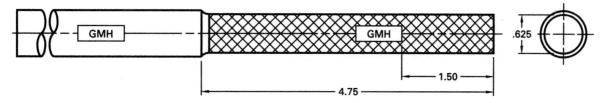

FIG. 2-2

12. Rough-turn the .625 diameter to .65 in.

13. Finish-turn the .625-in. diameter to the required length.

14. Knurl the diameter to the required length.

15. File a flat .25-in. wide × .75-in. long on the knurled section with a square file and stamp your initials on it (Fig. 2-2).

16. Rough-turn the A diameter to .562 in. This allows .06 in. for finish turning (Fig. 2-3).

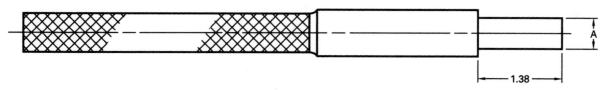

FIG. 2-3

17. Square-cut the corner and finish-turn the diameter to .500 in.

18. Cut the .12-in. groove to the dimensions shown in Fig. 2-4.

NOTE: Use extreme caution, as this diameter is very small and the handle can easily be bent.

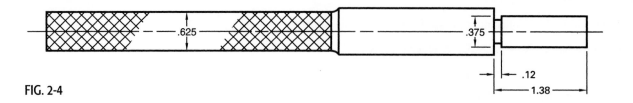

FIG. 2-4

19. From the dimensions in Fig. 2-5, calculate the taper per foot for the tapered section of the hammer.

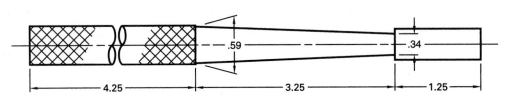

FIG. 2-5

20. Have your answer checked by the instructor.

21. Set the taper attachment and cut the taper.

22. Finish the .12-in. radius in the corner with a radius cutting tool (Fig. 2-6.)

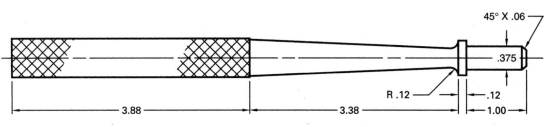

FIG. 2-6

23. Remove all lathe tool marks from the handle with a file.

24. Turn the 3.75-in. diameter and square the corner (Fig. 2-6).

25. Chamfer the .375-in. diameter end .06 in. × 45°.

26. Saw .25 in. off the knurled end of the handle.

27. Mount a three-jaw universal chuck on the lathe.

28. Protect the knurl from the chuck jaws by wrapping aluminum stock around the knurl and face to length (Fig. 2-6).

29. Set the compound rest to the required angle and finish the end of the handle (Fig. 2-7).

30. Protect the knurl from the vise jaws and use a .375-in. 16 UNC die to cut the thread on the end.

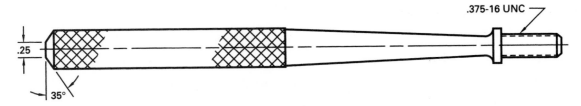

FIG. 2-7

Operations for Machining the Head

1. Cut off a piece of hot-rolled machine steel .75 × .75 × 4.31-in. long.

2. Stamp your initials on one end (Fig. 2-8).

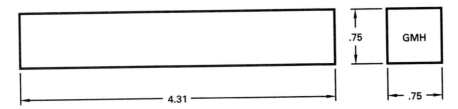

FIG. 2-8

3. Mill .03 in. from sides 1 and 2.

4. File one end square.

5. Apply layout dye to the two finished sides and lay out the .625-in. widths. Have the layout checked and then lightly prick-punch the lines.

6. Mill sides 3 and 4 to .625 in.

7. Lay out the tapered section as in Fig. 2–9. (Have the layout checked.)

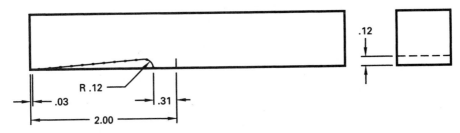

FIG. 2-9

8. With a round file, finish the .12-in. radius in the corner.

9. Set the hammer head in the mill vise with the layout line parallel to the top of the vise and cut to the layout line.

10. Lay out the .219-in. hole, as shown in Fig. 2-10.

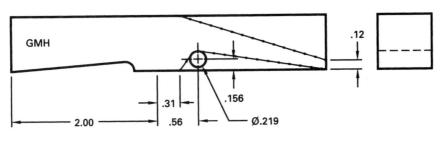

FIG. 2-10

11. Drill the .219-in. hole through the hammer head.

12. Lay out the remainder of the form shown in Fig. 2-10, have it checked by the instructor, and then lightly punch the lines.

13. Stamp your initials, as shown in Fig. 2-10.

14. Cut away the unwanted section on the contour bandsaw, staying at least .03 in. away from the layout lines.

15. Set the hammer head in the mill with the layout line parallel to the top of the vise, and cut to the layout lines (Fig. 2-11).

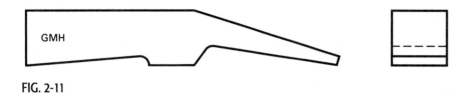

FIG. 2-11

16. Lay out the .312-in. hole, as shown in Fig. 2-12.

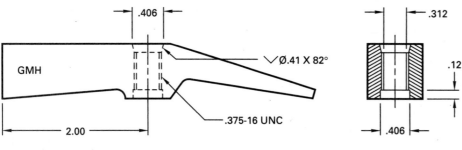

FIG. 2-12

17. Drill the .406-in. hole .12-in. deep.

18. Drill the .312-in. hole through the hammer head.

19. Countersink the .312-in. hole to .406-diameter (Fig. 2-12).

20. Thread the hole with a .375-16 UNC tap.

21. Lay out and file the beveled corners on the sides and face (Fig. 2-13).

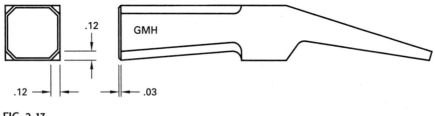

FIG. 2-13

22. Draw file and remove all machine marks.

23. Case-harden the hammer head.

24. Polish the hammer head and assemble it tightly with the handle.

25. Cut off the threaded section, leaving .06 in. past the hammer head for riveting.

26. Rivet the handle and file and polish the hammer.

TURNING AND THREADING EXERCISE - PROJECT #3

The turning and threading exercise is designed to provide students with the experience of machining a series of diameters to length. It is important to develop a passion for accuracy, and it should be everyone's goal to machine diameters to the exact size shown on the drawing. Even though the fractional length dimensions are given a tolerance of ± .016 in., it is always wise to produce work as close to the drawing size as possible.

OBJECTIVES

This project should provide the following learning experiences:

1. Turning diameters to accurate size and length.

2. Setting up a lathe and cutting standard threads to fit a test nut.

3. Machining short tapers using the compound rest.

4. Drilling and tapping a hole in the lathe.

Operations

1. Cut off one piece of hot-rolled machine steel 4.75-in. long.

2. Mount the steel in a three-jaw chuck having about 1.00 in. extending beyond the chuck jaws.

3. Face about .06 in. from this end and drill the center hole.

4. Turn the piece around in the chuck with about 1.00 in. beyond the chuck jaws, and face to 4.500 in. length.

5. Drill a center hole in this end.

6. Clean the tapers in the lathe spindle and tailstock and on the lathe centers.

7. Insert the centers into the lathe spindles with a sharp snap.

8. Fasten a suitable lathe dog on the part and mount the work between centers.

9. Rough-cut all diameters .03-in. larger than the size shown on the drawing, and all lengths .03 in. less than the sizes shown (Fig. 3-1).

Note: Diameter tolerances are ± .001 in.

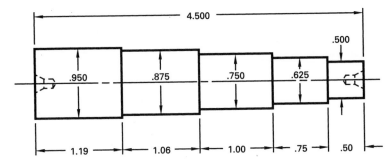

FIG. 3-1

10. Finish-cut all diameters and lengths to the size indicated on the drawing.

11. Use a parting or grooving tool to cut the first groove to the minor diameter of the .625-11 UNC thread (Fig. 3-2).

12. Cut the second groove to the minor diameter of the .750-in. 16 UNC thread.

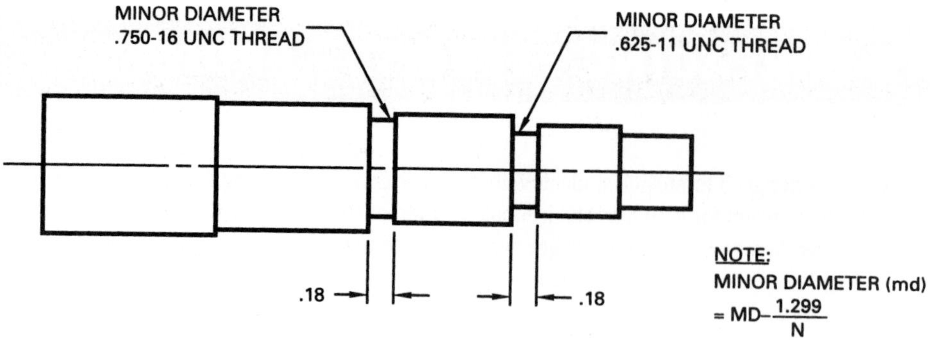
MINOR DIAMETER
.750-16 UNC THREAD
MINOR DIAMETER
.625-11 UNC THREAD
.18
.18
NOTE:
MINOR DIAMETER (md)
= MD–1.299/N

FIG. 3-2

13. Set up the lathe gearbox for 11 threads per inch.

14. Cut 11 threads per inch on the .625-in. diameter until the thread fits snugly into a test nut.

15. Set up the lathe gearbox for 16 threads per inch.

16. Cut 16 threads per inch on the .750-in. diameter until the thread fits snugly into a test nut.

17. Remove the part from the lathe and mount a three-jaw chuck.

18. Grip the part on the .875-in. diameter, having the large end extending beyond the chuck jaws.

19. Set the compound rest to 15° and cut the taper section on the end, using the compound rest feed-screw (Fig. 3-3).

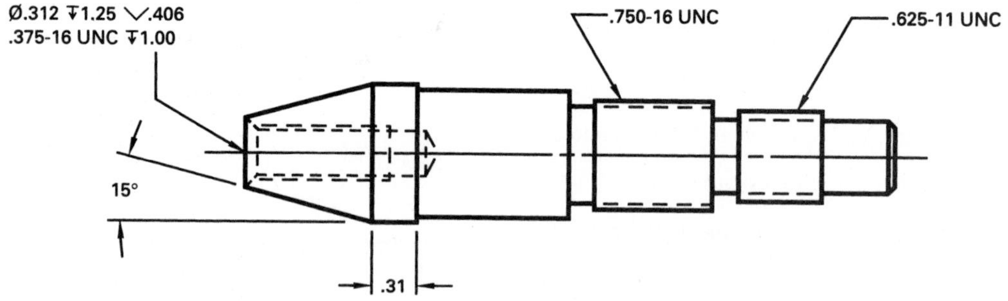
Ø.312 ↧1.25 ∨.406
.375-16 UNC ↧1.00
.750-16 UNC
.625-11 UNC
15°
.31

FIG. 3-3

20. Drill the tap drill size hole for the .375-16 UNC thread in the end of the part.

21. Countersink the end of the hole to .406-in. diameter.

22. Thread the .375-16 UNC hole: support the end of the tap with the tailstock center point.

Parallel clamps are used to fasten work to an angle plate or drill press table, and for holding two or more parts together. They are generally used where C-clamps are not suitable because of space restrictions, or to provide an easier access for the cutting tools.

OBJECTIVES

The parallel clamp project should provide the following learning experiences:

1. Machining two parts that should be similar and match on assembly.

2. Drilling and tapping accurate holes for the screws that align the jaws.

3. Cutting threads for the screws on long slender diameters.

4. Forming metal to produce the clip.

FIG. I-4

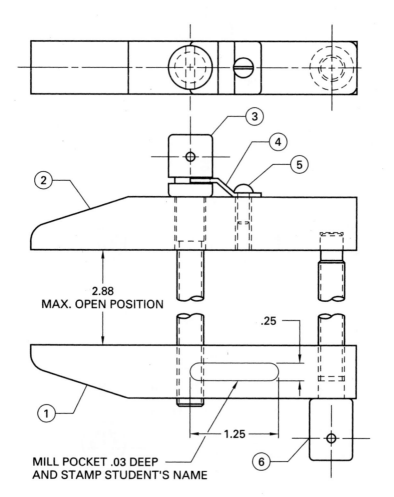

2.88
MAX. OPEN POSITION

.25

1.25

MILL POCKET .03 DEEP
AND STAMP STUDENT'S NAME

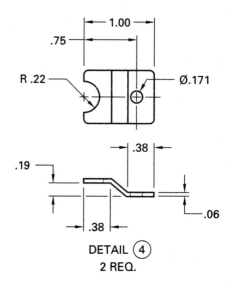

1.00

.75

R .22

Ø.171

.38

.19

.38

.06

DETAIL ④
2 REQ.

MATERIAL LIST			
DET. NO.	NO. REQ.	SIZE	MAT.
1	2	.75 X .75 X 4.75	C.R.S. PK. HRD.
2	2	.75 X .75 X 4.75	C.R.S. PK. HRD.
3	2	Ø.62 X 5.50	C.R.S.
4	2	.06 X .75 X 1.12	C.R.S.
5	2	8 - 32 X .38 RD. HD. SCREW	STK.
6	2	Ø.62 X 4.88	C.R.S.

PARALLEL CLAMPS
SHEET 1 OF 2
PRODUCT NUMBER 4

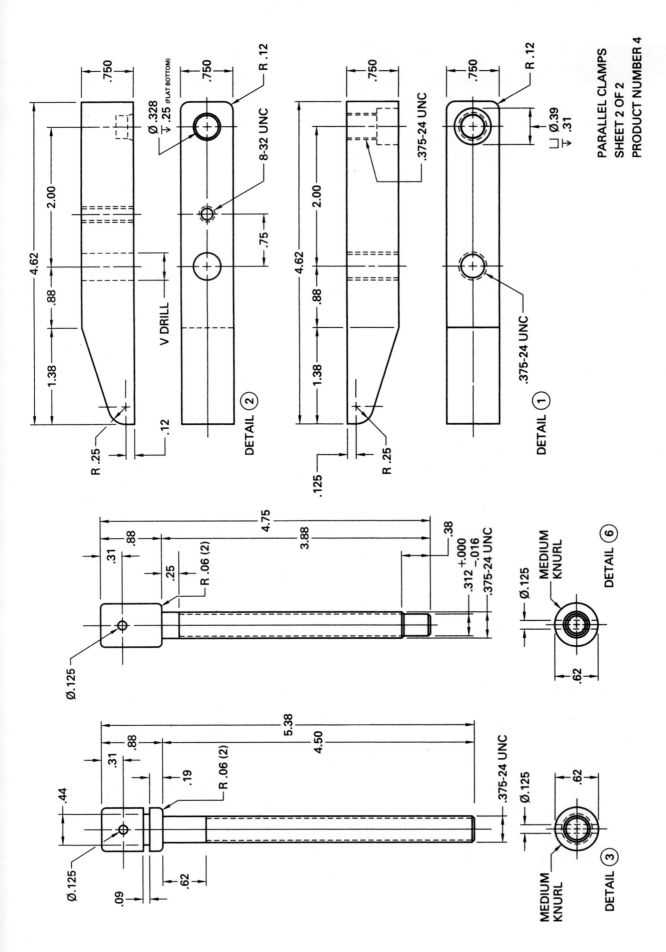

PARALLEL CLAMPS
SHEET 2 OF 2
PRODUCT NUMBER 4

.750

Ø .328
⤓ .25 (FLAT BOTTOM)

R .12

8-32 UNC

.750

V DRILL

.750

.375-24 UNC

.750

R .12

⌴ Ø .39
⤓ .31

.375-24 UNC

2.00

4.62

.88

1.38

R .25

.12

DETAIL ②

2.00

4.62

.88

1.38

R .25

.125

DETAIL ①

.75

4.75

.88

.31

3.88

.38

.25

R .06 (2)

.312 $^{+.000}_{-.016}$

.375-24 UNC

Ø.125

Ø.125

MEDIUM
KNURL

.62

DETAIL ⑥

5.38

.88

.31

4.50

.19

R .06 (2)

.44

Ø.125

.09

.62

.375-24 UNC

Ø.125

MEDIUM
KNURL

.62

DETAIL ③

The Jaws

1. Cut off two pieces of .75-in. square C.R.S. 4.75-in. long.

2. Clamp the two pieces, with their ends aligned, in a vertical milling machine vise with about .500 in. extending beyond the end of the vise.

3. Use an end mill to remove only enough material to square the end.

4. Remove the jaws from the vise and file off the burrs.

5. Replace the two jaws in the vise, align the squared ends, and machine the jaws to 4.625-in. length.

6. Lay out the hole locations and angular section on the Detail 1 jaw.

7. Lay out only the angular section on Detail 2.

8. Center-drill both hole locations in the Detail 1 jaw and then drill these holes through the jaw with a letter Q (.332) drill (tap drill for .375-24 UNC thread).

9. Countersink both holes to .406-in. diameter.

   NOTE: It is good practice to countersink all holes drilled to remove the burrs and sharp edges.

10. Clamp both jaws together, with the drilled jaw on top.
    • Assemble the jaws so that the angular surface of each jaw is facing out (see assembly drawing).
    • Be sure that the sides and ends of the jaws are exactly aligned and will not move in the spotting operation.

11. Use a letter Q drill to spot the holes approximately .12-in. deep in the Detail 2 jaw.

12. Disassemble the two jaws.

13. On the top of Detail 1 jaw, use a .39-in. drill to counterbore the end hole .31-in. deep.

14. Use a .375-24 UNC tap to cut the thread in each hole.

   NOTE: Be sure the holes are tapped square with the top of the vise jaw; otherwise, the screws will not align properly when the unit is assembled.

15. Use a letter V (.377-in.) drill to drill the spotted hole in the center of Detail 2 through the jaw.

16. Use a .328-in. drill to drill the end hole .25-in. deep, measuring from the drill point.

17. With a flat-bottom drill, square the bottom of the hole.

18. Lay out, drill, and tap the #8-32 UNC hole.

19. Machine the angular surface on each jaw by setting the layout line parallel to the top of the vise jaw.

20. File the .25-in. radius on each jaw at the end of the angular section.

21. Assemble both jaws with .375-24 UNC screws.

22. File a .125-in. radius on both sides of the right-hand end of the jaws.

23. Mill a pocket on each jaw for your name.

24. Case-harden the jaws.

The Screws

1. Cut off one piece of hot-rolled machine steel .75-in. diameter × 5.88-in. long for Detail 3.

2. Cut off one piece of hot-rolled machine steel .75-in. diameter × 5.25-in. long for Detail 6.

3. Drill center holes in the ends of each piece.

   NOTE: These center holes will be removed later.

4. Machine the .625-in. diameter to size on each screw.

5. Reverse the parts in the lathe and machine the .375-in. diameter to length.
   - For Detail 3 (longer screw), the .375-in. diameter should be 4.75-in. long; .25-in. will be cut off, removing the center hole, later.
   - For Detail 6 (shorter screw) the .625-in. diameter should be 4.12-in. long.

6. Cut the .09-in.-wide groove in Detail 3 to .44-in. diameter.

7. On Detail 6, cut the .312-in. diameter, .625-in. long (.375 + .25 in. for the center hole).

8. Cut the .375-24 UNC thread on both screws to the lengths shown.

9. Grip each screw in a three-jaw chuck and remove the center holes by facing to the correct lengths.

10. Radius all edges shown on the drawing with a file.

11. Lay out, drill, and ream the .125-in. holes in the large end of each screw.

The Clip (Detail 4)

1. Cut off one piece of C.R.S. .06 × .75 × 1.25-in. long.

2. Square the right-hand end with a file.

3. Lay out the position of both holes from the right-hand end of the part.

4. Hold the clip with vise grips and place it on a block of wood for drilling the holes.

5. Drill the .171-in. and .44-in. holes through the clip.

6. Saw and finish the left-hand end to 1.000-in. length.

7. Remove all burrs and sharp corners.

8. Lay out the lengths from both ends where the clip is to be bent.

9. Place two .19-in. shims on opposite sides of the clip where the bends are to be made.

10. The clip will be formed to shape when this unit is tightened in a bench vise.

Assembling the Clamp

1. Assemble the clamp by inserting the correct screws in each jaw (see the assembly drawing).

2. Fasten the clip into place on Detail 6 with a #8-32 × .38-in. UNC long screw.

3. Adjust the screws to close the clamp jaws; they should be aligned with each other.

THREADING EXERCISE - PROJECT #5

In the machine tool trade, many different types of threads are used for various applications of fastening parts together, measuring tools and devices, moving materials and machine slides, and increasing force. The threading exercise is designed to provide experience in cutting various types of threads.

OBJECTIVES

The threading exercise should provide the following learning experiences:

1. The calculations for tool dimensions and shape for various threads.
2. The lathe setup required for each thread.
3. The use of formulas for calculating thread dimensions.
4. The one- and three-wire methods of measuring threads for accuracy.

Operations

1. Cut off one piece of hot-rolled machine steel 1.25-in. diameter × 10.12-in. long.
2. Face the part to 10.00-in. long.
3. Drill center holes in each end.
4. Mount the part between lathe centers.
5. Rough-turn all diameters .03-in. oversize and .03-in. shorter in length.
6. Finish-turn the 1.125-in. diameter (center section to be knurled) 3.00-in. long.
7. Use a medium knurl to produce the pattern on this section.

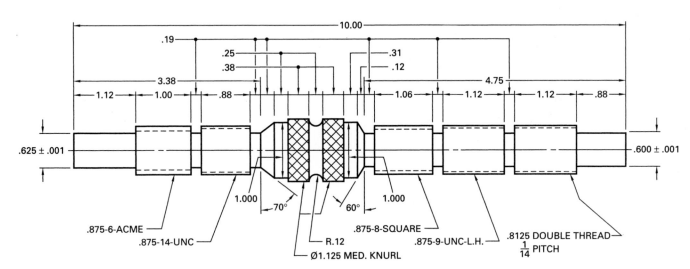

FIG. 5-1

8. Finish-turn all diameters to size and the correct length.
9. Calculate the minor diameter for the five different threads.
10. Use a grooving tool to cut the .19-in. groove at the end of each thread to its minor diameter.
11. Use the side of a 60° tool to chamfer both edges of each section to be threaded to its minor diameter.
12. Set up the lathe for cutting 60° UNC threads.
13. Calculate and grind the width of the 60° tool point for each UNC thread to be cut.

14. Calculate the three-wire measurement for each 60° thread to be cut.

15. Cut the threads to size, measuring the accuracy with the three-wire system or a thread micrometer.

16. Set up the lathe for cutting the .875-in.-square thread. (The compound rest should be set to 90° to provide side movement for the finish cuts.)

17. Calculate the width of the square thread and grind a cutting tool to .003-in. less than the calculated size.

18. Set up and square the threading tool.

19. Cut the thread to depth by taking successive cuts using the crossfeed screw micrometer collar.

20. After the thread has been cut to depth, feed the compound rest .0005 in. for each cut until the thread is the correct width.

21. Set up the lathe for cutting the .875-6 Acme thread.

22. Calculate and grind the 29° tool to shape and the correct point width.

23. Grind a grooving tool .005-in. smaller than the calculated Acme tool point.

24. Rough-cut the Acme thread to the correct depth with the grooving tool by taking successive cuts with the crossfeed handle.

25. Set up the Acme tool for thread cutting with its edge about .06 in. to the right of the edge of the groove.

    • This should bring the Acme tool point close to the center of the rough-cut groove when it reaches the bottom.

26. Cut the Acme thread to depth using the compound rest handle.

27. Measure the Acme thread for accuracy using the one-wire method. The wire diameter = .4872 × pitch of thread.

28. Remove all burrs and sharp corners.

Many times in the machine tool trade it is necessary to accurately drill and ream holes to specific locations to fit other parts on assembly. The easiest and a fairly accurate method of locating, drilling, reaming, and performing drill-related operations is on a vertical mill. To produce accurate work, the machine spindle must be aligned at 90° to the table and the part clamped or held in a vise parallel to the table travel.

OBJECTIVES

This milling and drilling project should provide the following learning experiences:

1. Accurate layout with a vernier height gage.

2. Drilling, reaming, tapping, countersinking, and counterboring on a vertical mill.

3. Milling slots and angular surfaces.

4. Setting work to an accurate angle using a sine bar.

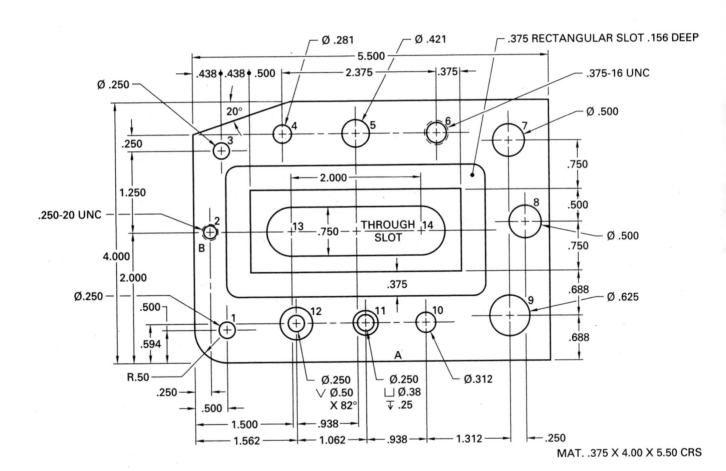

MAT. .375 X 4.00 X 5.50 CRS

Operations

1. Cut off one piece of C.R.S. .375 × 4.00 × 5.75-in. long.
   - Since the material is C.R.S., the thickness and width are to size.

2. Machine the two ends square and to 5.500-in. long.

3. Apply layout dye to one of the large surfaces.

4. Place the part on a surface plate with side A down and clamp it against an angle plate.

5. With a vernier height gage, scribe all horizontal lines for the holes (#1 to #13) and the slots on the part.
   - To avoid confusion because of the many lines, scribe the lines only slightly longer than required.

6. Place edge B on the surface plate and clamp the part against the angle plate.

7. Using a vernier height gage, scribe all the vertical lines for the holes and slots on the part.

8. Check all layout lines for accuracy.

9. Lightly prick-punch all hole locations, using a magnifying glass to check the accuracy.

10. Prick-punch the center of the radius at both ends of the .750-in. slot.

11. With a divider, scribe the radius for both ends of the .750-in. slot and the .25-in. radius on the lower left-hand corner.

12. Lay out the 20° angular section (top left-hand corner).

13. Lightly prick-punch all layout lines.

14. Set the part on parallels in the vertical milling machine vise that has been aligned parallel to the machine table.
    - Be sure the parallels are set so that they will not interfere with the drilling or milling operations.

15. Use an edge finder and locate the spindle center on edges A and B.
    - Turn the crossfeed and table handwheels in a counterclockwise direction when setting the spindle on an edge to remove the backlash.

16. Set the crossfeed and table graduated collars to zero (0).

NOTE: Do not move either handle when setting the collars.

17. Mount a center drill in a drill chuck in the machine spindle.

18. Make a **RECORD SHEET** to list the crossfeed and table graduated collar settings for all 13 holes as they are made.

    e.g., Hole 1   Crossfeed Setting   Table Setting
          ↓
          Hole 13

    - This record is important because it will be necessary to return to the same locations more than once.
    - All the final moves to a hole location must be made in a counterclockwise direction; otherwise, errors will occur because of backlash.

19. Lightly spot each hole location, check it for accuracy with a rule, and then center drill to about the middle of the taper portion of the drill.

20. Drill all holes to the size indicated on the drawing.

21. Drill all holes to be reamed:
    - .016-in. smaller for holes up to .500-in. diameter.
    - .03-in. smaller for holes over .500-in. diameter.

22. Slightly chamfer all holes to remove the burrs and sharp edges.

23. Countersink and counterbore the two holes to size.

24. Drill the holes to be tapped to the tap drill size.

25. Countersink the holes to be tapped .03-in. larger than the tap diameter.

26. Support the tap with a stub center held in the drill chuck and tap the holes through the plate.

27. Drill two .718-in. holes at locations 13 and 14.

28. Ream the two .750-in. diameter holes.

29. Use a .625-in. end mill to cut the .750-in. slot to size through the part.

30. Use a .375-in. end mill to cut the rectangular slot .156-in. deep.

31. On a contour bandsaw, saw the .50-in. radius corner and the 20° angular surface to within .03 in. of the layout line.

32. Set up the part on a sine bar set to 20° and clamp it to an angle plate.

33. Surface grind the 20° angle to size.

34. File the .50-in. radius to size on a contour bandsaw or a bench filer.

These 1 × 2 × 3 blocks are used for accurately setting up work on surface grinders, milling machines, etc., and in various inspection operations. Material: Two pieces of machine steel, 1.25 × 2.25 × 3.25-in. (SAE 1020 or AISI C 1020).

FIG. I-7

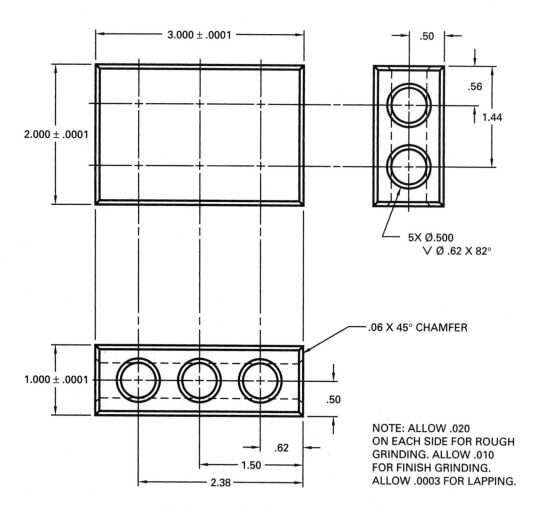

3.000 ± .0001

2.000 ± .0001

.50

.56

1.44

5X Ø.500
∨ Ø .62 X 82°

.06 X 45° CHAMFER

1.000 ± .0001

.50

.62

1.50

2.38

NOTE: ALLOW .020
ON EACH SIDE FOR ROUGH
GRINDING. ALLOW .010
FOR FINISH GRINDING.
ALLOW .0003 FOR LAPPING.

REQUIRED: TWO PIECES OF MACHINE STEEL (S.A.E. 1020).
CARBURIZE AND HARDEN. GRIND ALL SURFACES.

Machining Operations

1. Machine all six surfaces on a milling machine, leaving about .015 to .020 in. on each side for grinding before hardening the blocks.

2. Chamfer all edges .06 × 45° on a vertical milling machine with an end mill, or on a horizontal milling machine with a 45° cutter (Figs. 7-1 and 7-2).

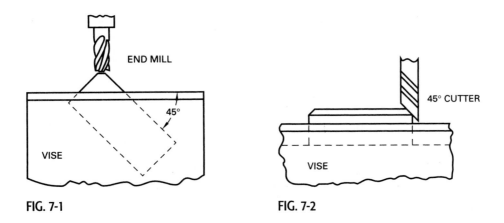

FIG. 7-1   FIG. 7-2

3. Rough-grind all six sides to within .010-in. of final dimensions before hardening.

> **CAUTION: Always be sure the magnetic chuck switch is on before starting the grinder.**

4. Lay out the hole locations for the drilling operation, as shown on the drawing.

5. If a vertical milling machine is available, lay out, center-drill, drill, and ream all the holes, or, if necessary, use a drill press.

6. Countersink all holes to .62-in. diameter with an 82° countersink.

7. Blocks should be carburized, hardened, and stress-relieved if machine steel is used. If tool steel is used, harden and temper according to specifications for the particular steel as recommended by the steel manufacturer.

> **CAUTION: After the blocks have been hardened, allow them to cool before grinding. If the blocks have warped during the hardening process, level the pieces with paper shims or tin foil so that they sit flat on the magnetic chuck.**

8. Rough-grind the blocks on all six sides, starting with the 2.000-in. × 3.000-in. surfaces (Fig. 7-3). Be sure to leave .003 in. on each surface for finish grinding.

9. Use an angle plate and grind the 1.000-in. × 3.000-in. surface square with the 2.000 in. × 3.000-in. surface (Fig. 7-4).

10. Set the ground 1.000-in. × 3.000-in. surface on a clean magnetic chuck.

> **CAUTION: When grinding a piece of work whose height is greater than its base, always use back-up blocks to prevent the work from being pulled away from the magnetic chuck (Fig. 7-5).**

11. Grind the 1.000-in. × 3.000-in. surface to size leaving the .003 in. finishing grinding allowance.

12. Grind the 1.000-in. × 3.000-in. surface with the angle plate set on end (Fig. 7-6).

13. After all sides have been rough-ground to within .003 in. of the finished size, change to a finish-grinding wheel.

14. Finish-grind all surfaces to within ±.0003 in. of final size. This will allow the blocks to be lapped to size.

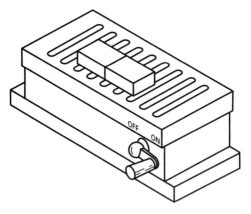

FIG. 7-3

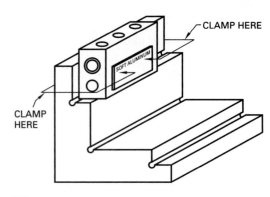

FIG. 7-4

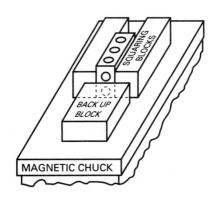

FIG. 7-5

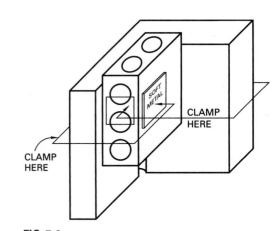

FIG. 7-6

15. After the grinding operation, lap the blocks to finished size on a serrated cast iron lapping plate (Fig. 7-7).

16. Use a fine oil stone to remove any burrs along the chamfered edges of the setup blocks.

17. If the chamfered surfaces have to be ground, clamp the block against an angle plate at a 45° angle (Fig. 7-8).

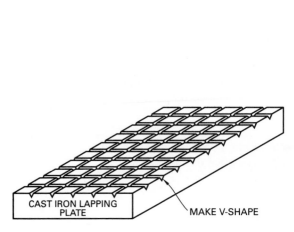

FIG. 7-7

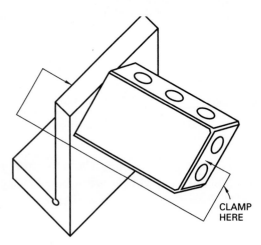

FIG. 7-8

DRILLING AND TAPPING FIXTURE - PROJECT #8

This handy fixture for the model maker or home craftsman can be used for drilling, tapping, or coil winding. As a project for machine shop students, it brings in the principles of cylindrical and internal grinding for producing a close fit between mating parts.

Material and Parts List

Part No.	No.	Mat.	Description
1 (A) & (C)	2	M.S	Drill stand, .625 × 3.00 × 4.25
1 (B)	1	M.S.	Drill Stand, .625 × 3.00 × 7.00
1 (D)	1	M.S.	Bed plate, .625 × 3.00 × 9.00
2	1	Standard	Chuck
3	1	M.S.	Bushing, 1.38 dia. × 1.00
4	1	M.S. or T.S.	Shaft, .75 dia. × 5.50
5	1	M.S.	Flywheel, 2.50 dia. × 1.00
6	1	Standard	Socket setscrew, .250 – 20 × .75
7	1	C.R.S.	Handle, .250 × 2.50
8	4	Standard	Flat-head capscrews .250 – 20 UNC × 1.000

Fixture Stand - Detail 1

1. The first step in making this fixture is the preparation of the drill stand and bed plate. The three pieces of the fixture stand can be fastened together with Allen socket-head setscrews or welded together.

2. Machine parts A, B, C, and D to size, leaving them .010 in. oversize for grinding.

3. Remove the machine marks by grinding parts A, B, C, and D to size.

4. Clamp parts A and B, and B and C to an angle plate for drilling holes for the two screws that will be used to fasten them together as one unit.

5. Drill and tap holes for the .312-18 UNC Allen Screws.

6. Fasten parts A and C to post B.

7. Fasten the stand to the base plate D with four .250-20 UNC flat-head machine screws.

8. Drill and ream the .875-in.-diameter hole for the bushing in part A.

9. Drill the .375-in.-diameter clearance hole in part D.

   NOTE: This hole must be in line with the .875-in. hole in part A.

Spindle - Detail 4

10. Turn the shaft to a diameter of .020 to .025-in. oversize, with the exception of the end to be threaded or tapered. Leave the threaded or taper end about .12-in. oversize in diameter and .06-in. in length (Fig. 8-1).

11. Carburize the spindle but do not harden it.

12. Turn the end to be threaded or tapered about .020 to .025-in. oversize.

13. Cut the spindle to length and, if necessary, recenter the end (Fig. 8-2).

14. Reheat the spindle to the hardening temperature and quickly quench it in a cooling medium.

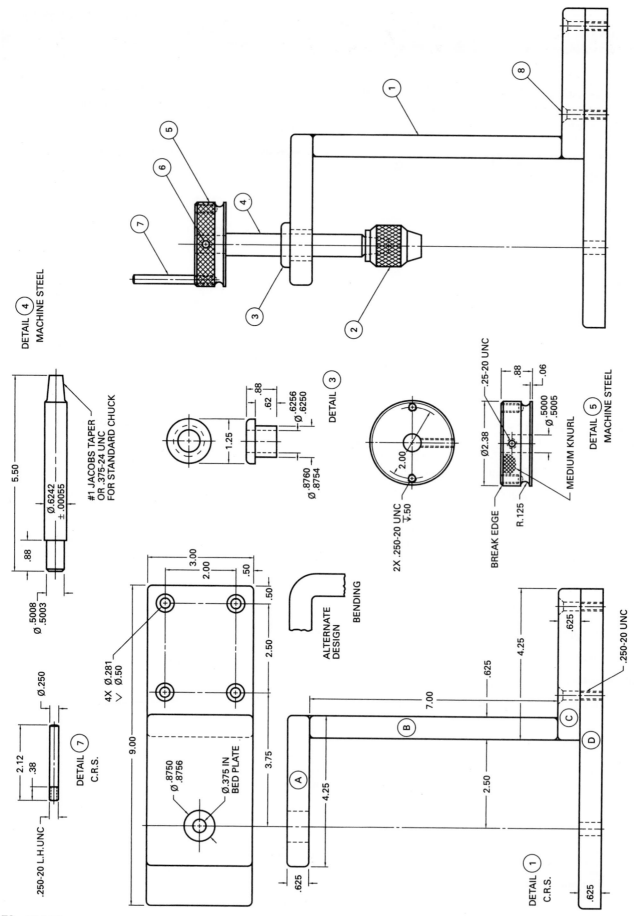

DETAIL 4
MACHINE STEEL

#1 JACOBS TAPER
OR .375-24 UNC
FOR STANDARD CHUCK

5.50

Ø.6242
±.00055

.88

Ø.5008
.5003

DETAIL 3

.88
.62

Ø.6256
.6250

1.25

Ø.8760
.8754

2X .250-20 UNC
↧.50

2.00

.25-20 UNC

.88

.06

Ø2.38

Ø.5000
.5005

MEDIUM KNURL

BREAK EDGE

R.125

DETAIL 5
MACHINE STEEL

Ø.250

2.12
.38

DETAIL 7
C.R.S.

.250-20 L.H. UNC

3.00
2.00
.50

.50

4X Ø.281
∨ Ø.50

2.50

9.00

3.75

Ø.8750
.8756

Ø.375 IN
BED PLATE

ALTERNATE
DESIGN

BENDING

A

.625

4.25

7.00

.625

B

C

2.50

D

4.25

.625

.250-20 UNC

.625

DETAIL 1
C.R.S.

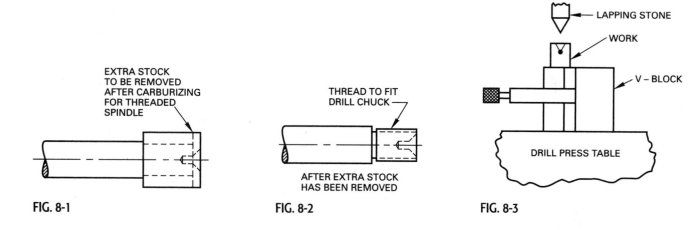

FIG. 8-1

FIG. 8-2

FIG. 8-3

15. Lap the center holes of the spindle to remove the scale caused by heat treating and to ensure accurate seating of the center points (Fig. 8-3).

16. Cut the thread or the taper, to suit the hole in the drill chuck, on the end left larger for carburizing and then later machined.

17. Grind the spindle to size using a cylindrical grinder, tool and cutter grinder, or toolpost grinder.

18. The body size should be ground for a medium running fit in the bushing hole.

19. The .500-in. diameter should be plunge-ground so that it is a press fit in Detail 5.

Bushing - Detail 3

20. Make the bushing out of machine or tool steel. The stock should be long enough so that it can be held in a four-jaw chuck (Fig. 8-4).

21. Turn the O.D. to .020 oversize and leave an extra .06-in. on the end to prevent a bellmouth on the hole (Fig. 8-5).

22. Center-drill the bushing and drill the hole .015 in. undersize to allow for grinding.

23. Harden the bushing.

24. Grind the I.D. with an internal grinder or a toolpost grinder (Fig. 8-6).

NOTE: The hole should be ground to provide a medium running fit for the spindle.

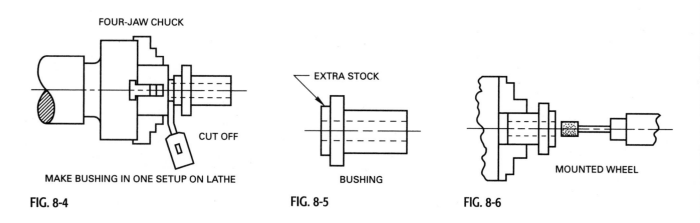

FIG. 8-4

FIG. 8-5

FIG. 8-6

25. Fasten the bushing in a V-block or a grooved angle plate and remove the excess stock on the ends by grinding (Fig. 8-7).

26. Press the bushing on the correct size mandrel for grinding the O.D. to size.

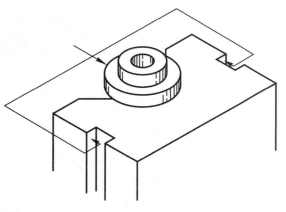

FIG. 8-7

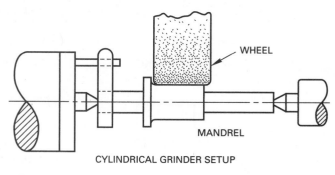

WHEEL

MANDREL

CYLINDRICAL GRINDER SETUP

FIG. 8-8

27. Grind the O.D. so that it is a press-fit between the bushing and the hole in part A.
   • If a cylindrical grinder is available, plunge-grind this short diameter (Fig. 8-8).
   • This diameter can also be ground using a toolpost grinder (Fig. 8-9).

28. Make all other parts required for the fixture according to the specifications on the print.

29. Assemble the parts to make a completed fixture.

30. The fixture can be motorized and used for light-drilling operations by using an old sewing machine motor with a foot-controlled variable speed device.
   • A round drive belt will fit the groove cut in the flywheel, Detail 5.

31. A pressure arm assembly (Fig. 8-10) can be made to provide a downfeed for drilling.
   • A small ball bearing, pressed into a hole in the point of the pressure arm, will reduce friction.
   • To provide an upfeed when drilling, place a medium gage spring around the spindle between the flywheel and the top of the bushing.

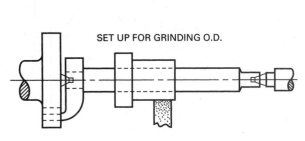

SET UP FOR GRINDING O.D.

PORTABLE GRINDER FASTENED IN TOOLPOST

FIG. 8-9

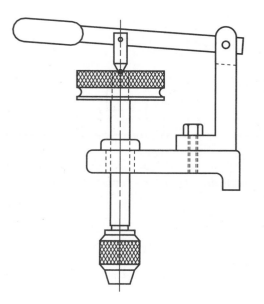

FIG. 8-10

5" SINE BAR - PROJECT #9

Sine bars are used wherever very accurate work is done, as in tool rooms or in experimental and inspection departments. They are used for laying out work, setting up work in machine tools, and checking the accuracy of finished work.

The accuracy required in producing this precision tool makes it an ideal project for advanced students. Completing this project will give the student an excellent conception of the accuracy demanded by the close tolerance requirements of industry.

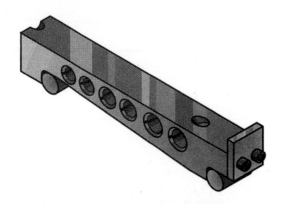

Material and Parts List			
Part	Number Required	Material	Description
1	1	T.S.	1.25 × 1.25 × 6.38 in., harden and temper
2	1	T.S.	.62 diameter × 3.00
3	2	Standard	8-32 UNC × 1.00 in. socket head screws
4	1	Gage Steel	.18 × .88 × 1.06 in., harden and temper
5	2	Standard	6-32 UNC × .50 in. socket head setscrews

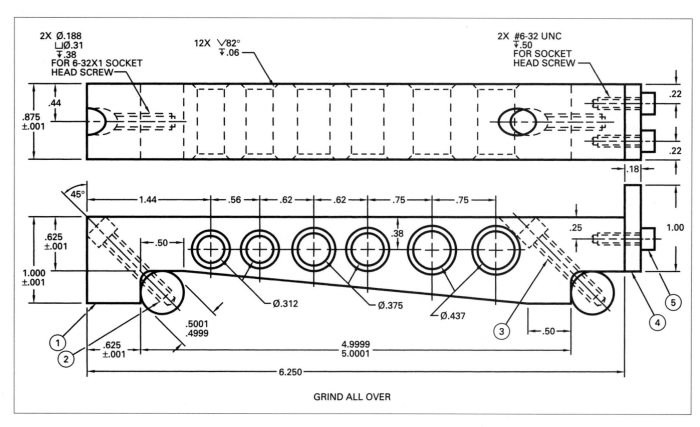

GRIND ALL OVER

Procedure

1. Machine all sides of the body of the sine bar (Part 1) .020 in. oversize on a vertical milling machine. All surfaces must be smooth and square with each other (Fig. 9-1).

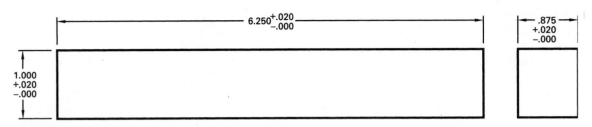

FIG. 9-1

2. Surface-grind all sides .010 in. oversize, using a 19A46-J5VG straight wheel (or equivalent).

> **CAUTION: Be sure magnetic chuck switch is closed before starting to grind.**

3. Use an angle plate when grinding the narrow sides and ends of the sine bar (Fig. 9-2).

> **CAUTION: When changing the position of the sine bar on the chuck, stop grinding wheel to avoid injury to the hands. Before mounting any work or accessories on a magnetic chuck, be sure it is thoroughly clean.**

4. Use a surface or height gage to lay out all dimensions. Prick witness marks along the lines to be cut (Fig. 9-3).

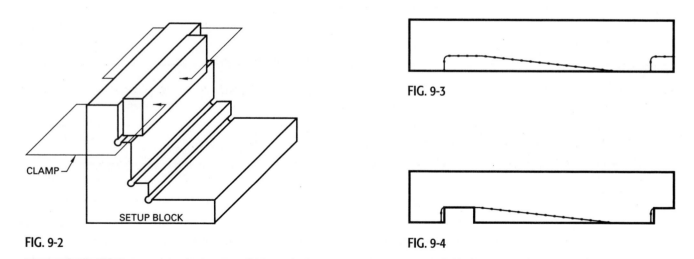

FIG. 9-2

FIG. 9-3

FIG. 9-4

5. Mill the ends of the offsets for .500-in. diameter rolls to within .015 in. of the finish marks (Fig. 9-4).

6. Mill the angular part of the sine bar body to the layout line.

7. Clamp the part at a 45° angle in a vise on a vertical milling machine.

8. Use a two-flute end mill, slightly larger than the head of the socket-head screw to be used, to mill a flat spot at each hole location (Fig. 9-5).

9. Center-drill the two hole locations.

10. Drill clearance holes for the #8-32 UNC screws through the sine bar with a #9 (.196 in.) diameter drill.

11. Counterbore the holes for the socket head setscrews deep enough to allow the head to seat below the surface of the sine bar (Fig. 9-6).

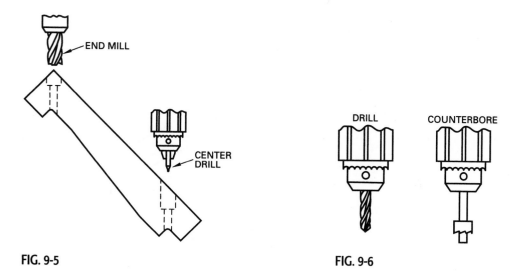

FIG. 9-5

FIG. 9-6

12. Use the holes in the end plate, Part 4, as a template and drill and tap the two #6-32 UNC holes on the end of the sine bar.

13. Locate, center-drill, drill, and countersink the holes on the side of the bar.

14. Check to be sure that all work has been done to the bar while it is still in the soft state.

15. Harden and temper the sine bar body, following the recommendations of the steel manufacturer or supplier.

16. After heat treating, the piece should be cleaned with abrasive cloth or by sandblasting to remove heat-treating scale.

17. Rough-grind 1.00 × 6.25-in. sides, allowing about .003 to .004 in. on each side for finish grinding.

18. Clamp the piece to the angle plate, again as in Fig. 9-2, and rough-grind the 6.250 × .875-in. side to within .003 to .004 in. for finished size.

19. Turn the work over on the magnetic chuck and grind the flat surfaces on the 1.00-in. dimension, leaving .003 to .004 in. for finish grinding.

20. Clamp the bar against an angle plate (Fig. 9-7 on p. 176) and grind the ends to within .005 to .006 in. of the finished size.

21. Square the sine bar with back-up rail on the grinder table and use back-up plates on both sides of the sine bar to prevent it from moving during grinding (Fig. 9-8 on p. 176).

CAUTION: Be sure that the sine bar touches both positive and negative poles on the magnetic chuck.

22. The edge of the wheel should be rounded to about a .12-in. radius (Fig. 9-9 on p. 176).

23. The angular part of the bar can be ground to give it a good appearance.

24. After the bar has been rough-ground all over, set it aside for a few days to allow stresses in the bar to equalize.

25. Finish-grind the bar to within .0002 in. of final dimensions.

26. The 5.000-in. dimension is the most critical dimension on the sine bar.
    • To check this measurement, use a surface plate with gage blocks, indicator, height gage, and angle plate (Fig. 9-10 on p. 176).

Sine Bar Rolls

27. The .500-in. diameter rolls can be machined from a piece of .62-in.-diameter tool steel.

28. Center-drill the piece and turn it to .520 to .525 in., undercutting one end for dog drive.

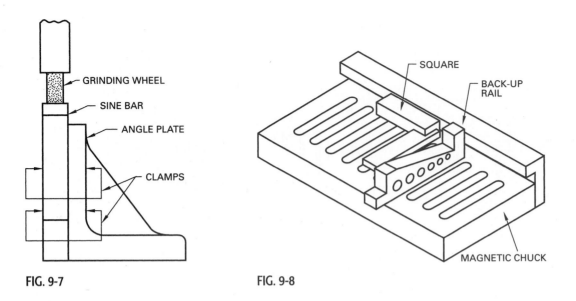

FIG. 9-7

FIG. 9-8

FIG. 9-9

FIG. 9-10

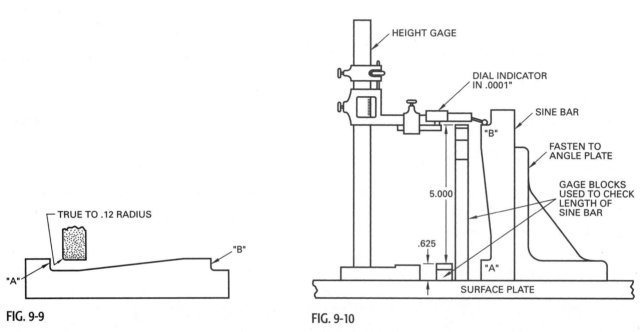

29. Undercut a small section between the two rolls, leaving enough stock on the length dimensions to remove the end center holes in the final step (Fig. 9-11).

30. Drill and tap the #8-32 UNC holes according to the drawing.

CAUTION: Be sure the holes are drilled exactly in the center of the diameter to ensure they will align with the holes in the sine bar on assembly.

31. Harden the rolls according to the steelmaker's recommendations.

32. After the rolls have been heat-treated, clean them with abrasive cloth.

33. Clean the center holes with a center lap mounted on a drill press to be sure they are true.

34. The rolls should be ground to .5000 in. ± 0.0001 in., either in a cylindrical grinder (Fig. 9-12), or in a tool and cutter grinder equipped with headstock and footstock centers.

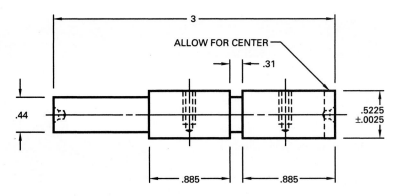

FIG. 9-11

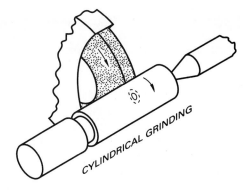

FIG. 9-12

35. After the rolls have been ground, clamp them in a small vise or V-block and cut them apart with a .12-in.-thick abrasive cutoff wheel (Fig. 9-13).

36. Set up the rolls in a V-block to grind both ends to a final size of .870 in. (Fig. 9-14).

End Plate (Detail 4)

37. Drill and machine a piece of tool steel gage stock, .18 × .875 × 1.06 in., as shown on the drawing. Chamfer all sharp edges to remove the burrs.

38. Harden and temper the end piece according to the steelmaker's recommendations.

39. Grind to size as shown, keeping the width to .870 in. to prevent end plate from extending beyond the sides of the sine bar when assembled.

Assembly

40. After all parts have been finished, assemble the sine bar and check the distance between the two rolls for accuracy.

41. If the distance between centers of the cylinders is not exactly 5.000 in., remove one roll and regrind the edge the amount of difference in measurement.

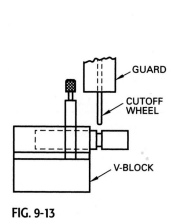

FIG. 9-13

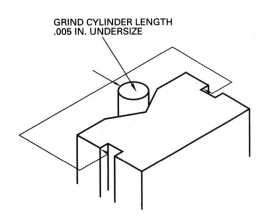

FIG. 9-14

TOOLMAKER'S VISE - PROJECT #10

Toolmakers and machinists can find a small precision vise a very useful and handy tool to have at their benches. It can make an ideal addition to any toolbox because of its many uses in the machine tool trade. The main body and the movable jaw may be made of machine steel or tool steel if the proper care is used in the heat treating operation.

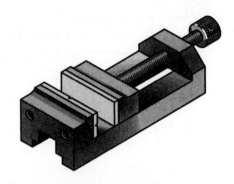

Material and Parts List

Part	Number Required	Material	Description (inches)
1	1	M.S. or T.S.	2.00 × 2.50 × 5.50
2	2	Standard	Socket-head cap screw .250-20 UNC × .75
3	2	T.S. gage stock	.312 × 1.50 × 2.50
4	1	M.S. or T.S.	1.25 × 1.75 × 2.50
5	2	Standard	Socket-head cap screw .250-20 UNC × 1.00
6	2	Standard	Flat-head socket head cap screw .250-20 UNC × .50
7	1	Standard	Socket-head set screw .250-20 UNC × .38
8	1	T.S. gage stock	.250 × 1.250 × 2.000
9	1	Standard	Socket-head cap screw .375-16 UNC × 4.25
10	1	Bronze	Bushing (see drawing)
11	1	Stainless Steel	Knurled handle (see drawing)
12	1	Standard	Socket-head set screw #10-32 UNC × .25

Machining Operations

1. Machine all surfaces of the body of the vise, Detail 1, to rough O.D. using a milling machine.
   • Leave .020 in. grinding allowance (Fig. 10-1).

2. Rough-surface-grind the body to .010 in. oversize, leaving this material for finish-grinding after hardening.

CAUTION: Always be sure that the magnetic chuck switch is in the "ON" position before grinding.

3. Grind the two widest sides (1) and (2) parallel (Fig. 10-2).

4. Clamp the part to an angle plate (Fig. 10-3) to ensure squareness and grind side (3).

5. For grinding side (4), place side (3) against the magnetic chuck.

6. When grinding the two ends, (5) and (6), be sure work is securely fastened to an angle plate and that the sides already ground, (1) and (3), are square with the chuck (Fig. 10-4).

CAUTION: When grinding a piece of work whose height is greater than its base, always use holding blocks or an angle plate to support the part.

7. Apply layout dye to Detail 1, place the part on a surface plate, and lay out the vise body for the opening, guide slot, angular part, and the holes with a height gage (Fig. 10-5).

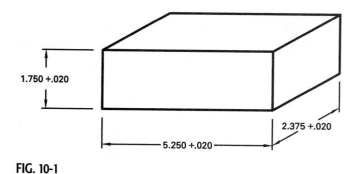

1.750 +.020

2.375 +.020

5.250 +.020

FIG. 10-1

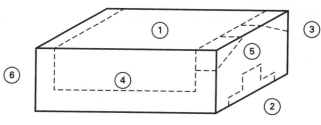

FIG. 10-2

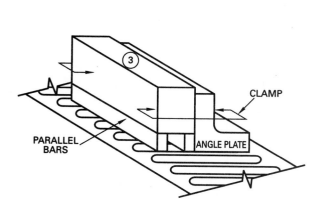

CLAMP

PARALLEL BARS

ANGLE PLATE

FIG. 10-3

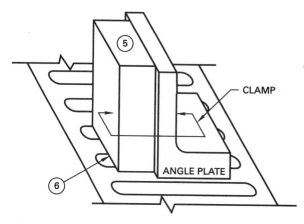

CLAMP

ANGLE PLATE

FIG. 10-4

8. The center part of the body and the angular corners on the right-hand end may be machined on a milling machine.

CAUTION: Always remove burrs and sharp corners produced by the machining operation with a file to prevent injury to hands or inaccuracies in the work.

9. Draw-file and polish all corners shown on the drawing.
   - If a small radius cutter is available, this operation can be done with an end mill or a corner rounding cutter on a milling machine (Fig. 10-6).

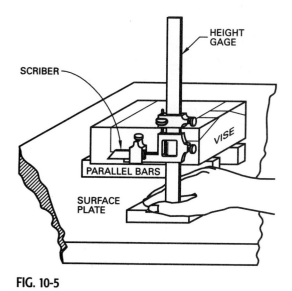

HEIGHT GAGE

SCRIBER

VISE

PARALLEL BARS

SURFACE PLATE

FIG. 10-5

END MILL GROUND TO PROPER RADIUS

CORNER ROUNDING CUTTER

A GRINDING WHEEL TRUED TO PROPER RADIUS ALSO CAN BE USED

VISE

FIG. 10-6

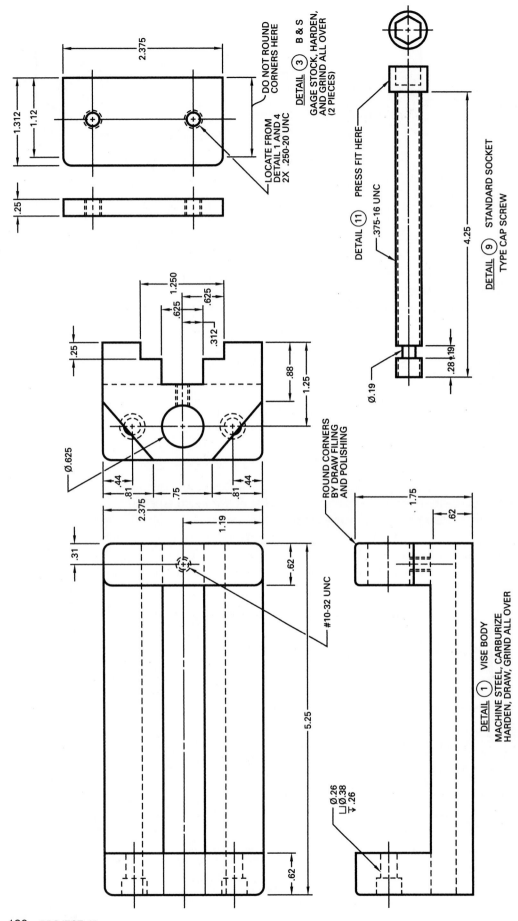

2.375

1.312

1.12

.25

DO NOT ROUND
CORNERS HERE

DETAIL ③ B & S

GAGE STOCK, HARDEN,
AND GRIND ALL OVER
(2 PIECES)

LOCATE FROM
DETAIL 1 AND 4
2X .250-20 UNC

PRESS FIT HERE

DETAIL ⑪

.375-16 UNC

Ø.19

4.25

.28 .19

DETAIL ⑨ STANDARD SOCKET
TYPE CAP SCREW

1.250

.625

.625

.312

.88

1.25

Ø.625

.44

.81

.75

.81

.44

2.375

1.19

.31

#10-32 UNC

ROUND CORNERS
BY DRAW FILING
AND POLISHING

1.75

.62

.62

5.25

Ø.26
⌴Ø.38
⯆.26

.62

DETAIL ① VISE BODY

MACHINE STEEL, CARBURIZE
HARDEN, DRAW, GRIND ALL OVER

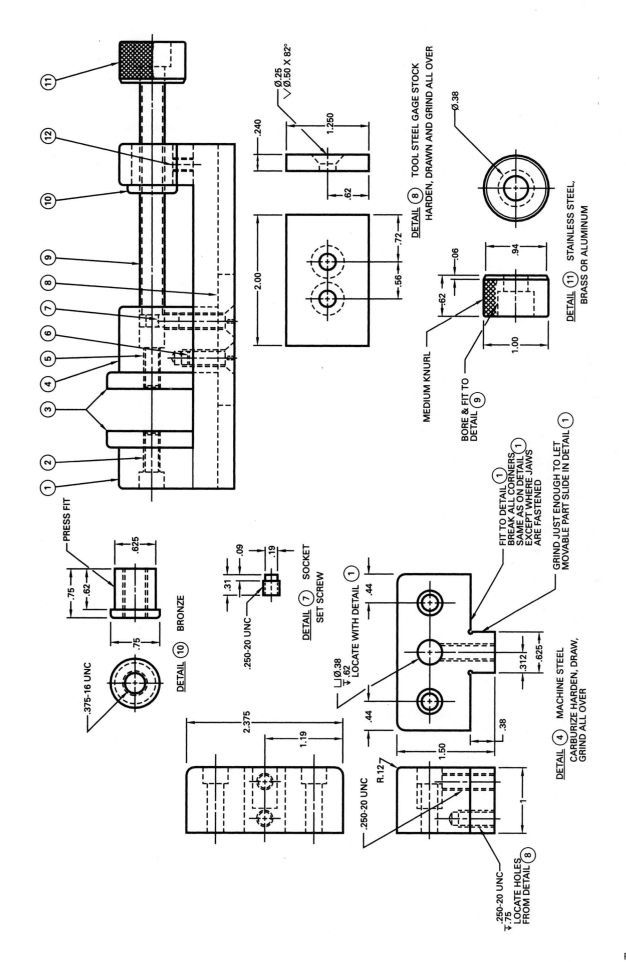

∅.25
∨∅.50 X 82°

1.250

.240

.62

2.00

.72

.56

DETAIL 8 TOOL STEEL GAGE STOCK
HARDEN, DRAWN AND GRIND ALL OVER

∅.38

STAINLESS STEEL,
DETAIL 11 BRASS OR ALUMINUM

MEDIUM KNURL

BORE & FIT TO
DETAIL 9

.94

.06

.62

1.00

PRESS FIT

.625

.75

.62

.75

.375-16 UNC

DETAIL 10 BRONZE

.31

.09

.19

.250-20 UNC

DETAIL 7 SOCKET
SET SCREW

FIT TO DETAIL 1
BREAK ALL CORNERS
SAME AS ON DETAIL 1
EXCEPT WHERE JAWS
ARE FASTENED

GRIND JUST ENOUGH TO LET
MOVABLE PART SLIDE IN DETAIL 1

⌴∅.38
▼.62
LOCATE WITH DETAIL 1

.44

.44

.312

.625

.38

1.50

DETAIL 4 MACHINE STEEL
CARBURIZE HARDEN, DRAW,
GRIND ALL OVER

2.375

1.19

R.127

.250-20 UNC

.250-20 UNC

▼.75
LOCATE HOLES
FROM DETAIL 8

1

10. Use a dial indicator to check the vise for squareness and then cut the guide slot with an end milling cutter (Fig. 10-7).

11. Drill and counterbore the two clearance holes for the .250-20 socket-head screws, Detail 2, in the solid jaw of the vise body (left side).

12. Drill and ream the .625-in. diameter hole in the right end of the body.

   NOTE: Clamp the vise body against an angle plate to be sure that the hole is square so that all vise parts will align on assembly.

13. Drill and tap the #10-32 UNC hole between the bottom slot and the .625-in. diameter reamed hole.

14. Slightly countersink all holes drilled, tapped, or reamed, to remove the sharp corners.

15. Harden the vise body, Detail 1, if desired, according to the manufacturer's specifications for the steel used.

16. Set up the work and grind the bottom of Detail 1 (Fig. 10-8).
   • Use back-up blocks to prevent the part from moving during the grinding operation.

17. Place the ground surface on the magnetic chuck and finish-grind the upper surfaces of the vise body.

18. Use two clamps to hold the vise body against an angle plate so that *one side* and *one edge* extend slightly beyond the surface of the angle plate (Fig. 10-9).

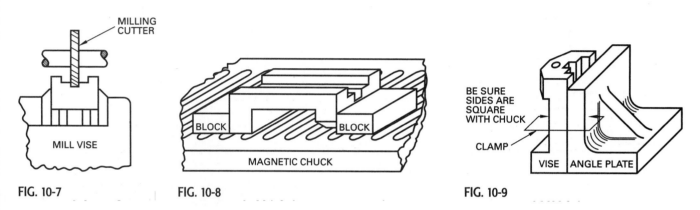

FIG. 10-7   FIG. 10-8   FIG. 10-9

   • Place the clamps so that they will not interfere with the grinding operation.

19. Finish-grind one end and then remove the angle plate and vise assembly from the magnetic chuck as a unit.
   DO NOT REMOVE ANY CLAMPS AT THIS TIME.

20. Set the angle plate on its edge and remove *only* one clamp.

21. Reset this clamp so that it will not interfere with the next operation of grinding the side.
   • Then loosen the other clamp and reclamp so that it will not interfere with grinding the side.

   CAUTION: Do not remove more than .010 in. from this surface to keep the bottom slot in the center of the vise body.

22. Place the assembly on the chuck, energize the chuck, and finish-grind the side of the vise body.

23. Remove the vise from the angle plate, place the ground side on the magnetic chuck, and grind the other side to 2.375-in. width.

24. Place the ground end on the magnetic chuck, clamp the vise body to an angle plate, and finish-grind the length to 5.250 in.

25. Use a square to set the vise body at right angles on the magnetic chuck (Fig. 10-10).

26. With a dish wheel, or one that has been relieved on the side, grind the center or inside surfaces of the vise.
   • Reverse the part on the magnetic chuck to grind the other vertical surface (Fig. 10-11).

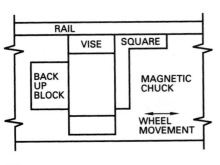

FIG. 10-10

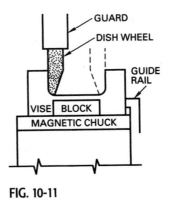

FIG. 10-11

27. After the inside of the vise body has been ground, turn the body base-side up, with the long side parallel to the guide rail.

28. Grind surfaces in the slot shown in Fig. 10-12, being sure that an equal amount is ground from each surface.
   • With a micrometer, measure from the edge of the slot to the outer edge of the body to *keep the slot in the center of the vise.*

29. Clamp the vise body to an angle plate and grind the angular part of the vise (Fig. 10-13).

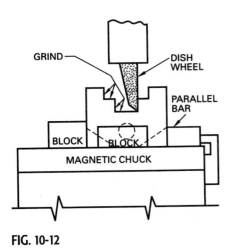

FIG. 10-12

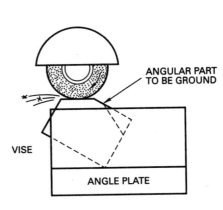

FIG. 10-13

Movable Jaw (Detail 4)

30. Machine the movable jaw, leaving .020 in. on all dimensions for grinding.

31. Finish-grind the outer surfaces of the movable jaw.

32. Clamp the part on an angle plate to grind the bottom surface and the tongue.

33. Grind the .625-in. dimension on the tongue to be a sliding fit in the .625-in. groove in Detail 1.

   NOTE: During grinding, be sure to use a depth micrometer to measure from each side to keep the tongue in the center of the part.

34. Round all edges shown on the drawing.

35. Use Detail 8 to locate, drill, and tap the two holes for the .250-20 UNC screws in the bottom face of the tongue.

36. Drill and counterbore the two holes for the .250-20 UNC screws that hold the vise jaw.

37. Insert the movable jaw, Detail 4, in the slot in Detail 1.

38. Clamp the jaw against the inside right side of the vise body.

39. Use the .625-in. diameter hole in the vise body (Detail 1) to locate the .375-in. counterbore hole in the movable jaw (Detail 4).

40. Remove the movable jaw, drill, and counterbore the .375-in. hole .62-in. deep.

41. Drill and tap the .250-20 UNC hole from the bottom into the .375-in. counterbored hole.

Other Parts

42. Make two vise jaws, Detail 3, from standard gage stock or tool steel. Allow .020 in. for finish-grinding after hardening.

43. Locate and drill the holes for the screws in one jaw by using the holes in Detail 1 as a template.
   • Use the holes in Detail 4 as a template for the other jaw.

44. Tap the holes in both jaws with a .250-20 UNC tap.

45. Harden and grind the jaws to size.

46. Make Detail 8 in the same manner as Detail 3.

47. The feedscrew for the vise, Detail 9, may be made from a standard socket-head cap screw or from a solid piece of stock.
   • The end of the screw should be a good press-fit into the knurled handle, Detail 11.

48. The outside diameter of the bronze bushing, Detail 10, must be a press-fit into the .625-in. hole in Detail 1.

49. Assemble all parts of the vise.

COMPUTER NUMERICAL CONTROL

The world of machine tools and manufacturing has undergone a continual change since the first numerical control (NC) machine tool was introduced in the late 1950s. These NC machines could be programmed to machine parts of consistent accuracy and repeatability, and they helped to increase the productivity of machines. By the mid 1990s, about 90 percent of the machine tools manufactured worldwide were equipped with some form of computer numerical control (CNC).

HOW NC WORKS

There is nothing magical or complex about numerical control. It is based on simple fundamentals that combine automatic measurement of distance with a preselected or programmed series of motions to move tool slides and start machine functions. Basically, numerical control is the operation of a machine by a series of coded instructions consisting of letters, numbers, and symbols. These coded instructions are converted into pulses of electrical current or other output signals that activate motors or other devices to operate the machine.

In a conventional machining operation, the machinist studies the part print; mounts and positions the part on the machine; selects the cutting tools, speeds, feeds, etc.; and turns the handles to move the machine slides to manufacture the part according to the sizes on the part print. In *numerical* control machining, the information about the part must be transferred to a manuscript or a computer, which in turn instructs electrical stepping motors and servomechanisms to machine the part automatically. The same basic machining principles used in conventional machining are used in CNC machining. However, the possibility of human error is always present in conventional machining; whereas once a CNC program has been properly prepared and checked, any number of parts can be produced quickly and to a very high degree of accuracy.

MANUFACTURING IN THE FUTURE

The future in the machine tool trade and manufacturing will require a working knowledge of CNC machine tools and their programming. Therefore, it is important for the student to take every opportunity to learn about CNC and the methods of programming the machine tools. Anyone with this knowledge will find that the opportunities in manufacturing are limited only by the initiative and effort put forth by each person.

The CNC projects in the next section are designed to provide experience in the programming of CNC bench-top teaching machines and standard lathes and mills. Learn from the programming techniques in each example and use them as references when starting to program other jobs yourself.

PROGRAM DISCLAIMER

Program examples are only for illustrations of a certain technique. Always refer to the programming and operations manuals for exact programming format.

MACHINING CENTER PROJECTS

Machining centers are basically milling machines equipped with computer numerical control (CNC) to guide the cutting tools and machine slides to automatically perform various machining operations. These machines are capable of greater precision and higher production rates than the standard milling machine.

The projects in this section are designed for two types of machines:

1. The CNC bench-top teaching mill that may have an EIA RS 274D or a Fanuc-compatible control.

2. The CNC standard machining center with a Fanuc-compatible control, commonly found in industry.

The basic programming for CNC machining centers is similar; however, some codes and minor programming procedures vary with the control found on each machine.

Multiple Hole Locations

Many times it is necessary to locate a series of holes on a part. Each hole can be programmed by providing the XY center location and the necessary codes to drill the hole through the part and bring the drill back out of the hole. To avoid repetitive programming for the drilling operation on each hole, a canned or fixed cycle can be included in the program. Canned or fixed cycles save valuable programming time and program memory for many types of repetitive machining operations.

OBJECTIVES

This multiple-hole-location project should provide the following learning experiences:

1. Using the **G01** and **G00** codes for moving the cutting tool to various locations and M-codes for some common machine functions.
2. Using canned or fixed cycles for a series of similar operations on a part.
3. Using the **G81** fixed cycle code for spotting and drilling holes.
4. Using the **G83** fixed peck drilling cycle code.

Program Notes

1. Material: .50 × 1.25 × 3.00 in. (CS 100).
2. High-speed steel tools required: .375-in. diameter spot drill, #10 (.193-in. diameter) drill, #16 (.177-in. diameter) drill, .186-in. diameter drill.
3. Program in the absolute mode.
4. Start programming at XY zero at the bottom left part edge.
5. Include all speeds, feeds, and codes required.
6. Use coolant when machining.
7. Use the G81 fixed drilling cycle for spot drilling and reaming; use the G83 fixed drilling cycle for the #10 and .172-in. holes.

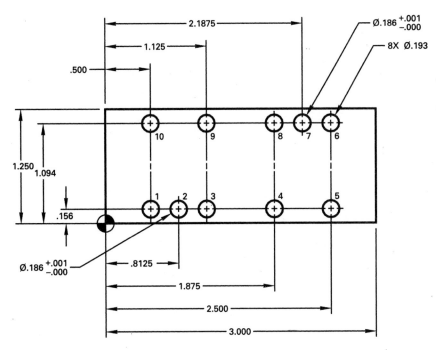

FIG. 11-1

The Program (Figure 11-1)

%

N010 G20 G90	Inch/Absolute
N020 T01 M06	Tool change .375-in. spot drill
N030 S1500 M03	Spindle on clockwise
N040 G54 G00 X.5 Y.156 Z.1 M08	Move to hole # 1/coolant on
N045 G43 Z1.0 H01	Tool Length Offset
N050 G81 X.5 Z-.15 R.1 F5.0	Hole #1 spot drill cycle
N060 X.8125	Hole #2
N070 X1.125	Hole #3
N080 X1.875	Hole #4
N090 X2.5	Hole #5
N100 Y1.094	Hole #6
N110 X2.1875	Hole #7
N120 X1.875	Hole #8
N130 X1.125	Hole #9
N140 X.5	Hole #10
N145 G80 G28 Z.1	Return to Machine Zero (Zaxis)
N150 G00 X6.0 Y0 T02 M06	Tool change # 10 drill
N160 S1500 M03	Spindle on clockwise
N170 G00 X.5 Y.156 Z.1 M08	Move to hole #1/coolant on
N175 G43 Z.1 H02	Tool Offset
N180 G83 X.5 Z-.600 R.1 Q.3 F5.0	Hole #1 G83 peck drill cycle
	Hole #3
N190 X1.125	Hole #4
N200 X1.875	Hole #5
N210 X2.5	Hole #6
N220 Y1.094	Hole #8
N230 X1.875	Hole #9
N240 X1.125	Hole #10
N250 X.5	Tool change #16 drill
N255 G80 G28 Z.1	
N260 G00 X6.0 Y0 T03 M06	Spindle on clockwise
N270 S1500 M03	
N280 G00 X.8125 Y.156 Z.1 M08	Move to hole #2/coolant on

The Program

N285 G43 Z1.0 H03	Tool Length Offset
N290 G83 X.8125 Z-.600 R.1 Q.3 F5.0	Hole #2 G83 peck drill cycle
N300 X2.1875	Hole #7
N305 G80 G28 Z.1	Return to Machine Zero (Zaxis)
N310 G00 X6.0 Y0 T04 M06	Tool change .186 diameter reamer
N320 S750 M03	Spindle on clockwise
N330 G00 X.8125 Y.156 Z.1 M08	Move to hole #2/coolant on
N340 G81 X.8125 Z-.600 R.1 F10.0	Hole #2 G81 fixed drill cycle
N350 X2.1875	Hole #7
N355 G80 628 Z.1	Return to Machine Zero (Zaxis)
N360 G00 X6.0 Y0 T1 M06	Tool change
N370 M30	End of program
%	Rewind/stop code

Three-Slot Milling

Sometimes it may be necessary to repeat a milling operation, such as a slot, in more than one place on a part. It is always necessary to provide the XY location of the start and end points of each slot in the program. It is also necessary to lower the cutter to the correct depth and then raise it above the part surface once the slot has been cut. A subroutine or macro, a miniprogram within the program, can be used to repeat the operation of raising and lowering the cutter as many times as necessary without having to program it again. In this project each slot is programmed using the codes for each step. All dimensions are in decimal inches.

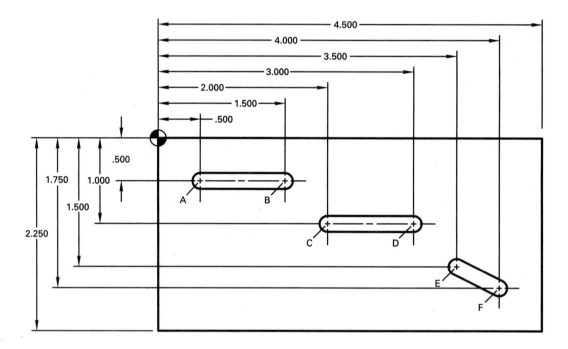

FIG. 12-1

OBJECTIVES

The three-slot milling project should provide the following learning experiences:

1. The use of **G00** code to move a cutter to location for the start of the slot.
2. The Z-axis movement for setting the cutter to depth and raising it above the work surface.
3. The **G01** code for milling the slots to length.

Program Notes

1. Material: Machine steel .50 × 2.25 × 4.50 in. (CS 100).
2. Use a high-speed steel 2-flute .375-in. diameter end mill.
3. Slot width is .375 in.; depth is .125 in.
4. Program in the absolute mode.
5. Programming starts at the XY zero (top left corner of the part).
6. Include all required speeds, feeds, and codes.
7. Use coolant when machining.

The Program (Figure 12-1)

%	
N010 G20 G90	Inch/absolute
N020 T01 M06	Tool change .375-in. end mill
N030 S1000 M03	Spindle on clockwise
N040 G54 G00 X.5 Y-.5 Z.1 M08	Move to point **A**/coolant on
N045 G43 Z1.0 H01	Tool Length Offset
N050 G01 Z-.125 F2.0	Feed to depth
N060 X1.5 F5.0	Machine to point **B**
N070 G00 Z.1	Move .100 above part surface
N080 X2.0 Y-1.0	Move to point **C**
N090 G01 Z-.125 F2.0	
N100 X3.0 F5.0	Machine to point **D**
N110 G00 Z.1	
N120 X3.5 Y-1.5	Move to point **E**
N130 G01 Z-.125 F2.0	
N140 X4.0 Y-1.75 F5.0	Machine to point **F**
N150 G00 Z.1	
N155 G28 Z.1	Return to Machine Zero (Zaxis)
N160 G00 X-2.0 Y0 T1 M06	Tool change
N170 M30	End of program
%	Rewind/stop code

Angular and Radii Milling

Angles and radii are fairly easy to produce on most machines equipped with a fairly recent control. To produce angles, the start point, the end point, and a vector feedrate must be programmed. Circular interpolation, available on most controls, allows the programmer to make a cutting tool follow any circular path ranging from a short arc to a full 360° circle. With some controls, the start and end point of the arc, the radius of the circle, and the coordinate location of the circle center must be supplied. Other controls will generate an arc if the arc radius and the end point of the arc are supplied. All dimensions are shown in decimal inches.

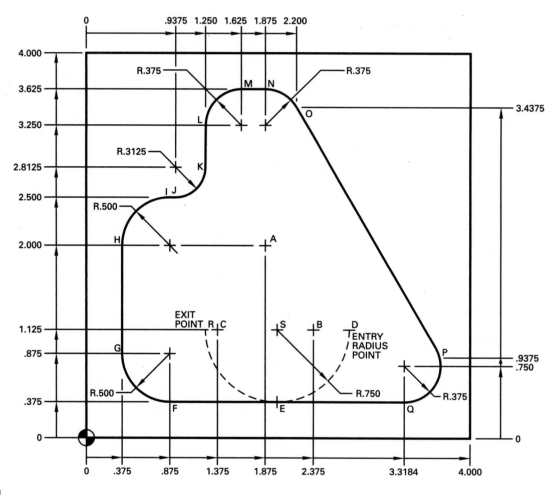

FIG. 13-1

OBJECTIVES

The angular and radii milling project should provide the following learning experiences:

1. The procedures for producing angular forms.
2. The procedure for producing radii by using the **G02** and **G03** circular interpolation codes.
3. The procedure for accurately blending a radius and an angle to produce a harmonious form.

Program Notes

1. Material: Aluminum .50 × 4.00 × 4.00 in. thick (CS 500).
2. Use a high-speed steel 2-flute .750-in.-diameter end mill.

3. Slot depth is .125 in.

4. Program-absolute mode.

5. All programming begins at the XY zero at the bottom left edge of the part.

6. Include all speeds, feeds, and codes required.

7. Use coolant when machining.

The Program (Standard Machining Center—Fanuc Compatible Control)

▲ CAUTION: Be sure this programming format suits your machine.

%	
O7755	Program number
N010 G0 G20 G90	Rapid, inch, absolute
N020 T01 M06	Tool change .750-in. end mill
N030 S2700 M03	Spindle on clockwise
N040 G54 X1.875 Y2.375 Z.1 M08	Move to .375 above point **A**/coolant on
N045 G43 Z1.0 H01	Tool Length Offset
N050 G01 Z-.125 F5.0	Feed to depth
N060 X2.375 Y1.125	Machine to point **B**
N070 X1.375	Machine to point **C**
N080 X1.625 Y2.375	Machine to .250 left of point **A** and .375 above
N090 G42 D01 G01 X2.75 Y1.125	Cutter compensation on, machine to point **D**
N100 G2 X2.0 Y.375 R-.75	Machine to point **E**
N110 G1 X.875	Machine to point **F**
N120 G2 X.375 Y.875 R.5	Machine to point **G**
N130 G01 Y2.0	Machine to point **H**
N140 G2 X.875 Y2.5 R.5	Machine to point **I**
N150 G1 X.9375	Machine to point **J**
N160 G3 X1.25 Y2.8125 R.3125	Machine to point **K**
N170 G1 Y3.25	Machine to point **L**
N180 G2 X1.625 Y3.625 R.375	Machine to point **M**
N190 G01 X1.875	Machine to point **N**
N200 G2 X2.2 Y3.4375 R.375	Machine to point **O**
N210 G1 X3.643 Y.9375	Machine to point **P**

The Program (Standard Machining Center—Fanuc Compatible Control)

▲ CAUTION: Be sure this programming format suits your machine.

N220 G2 X3.3184 Y.375 R.375	Machine to point **Q**
N230 G01 X2.0	Machine to point **E**
N240 G2 X1.25 Y1.125 R.75	Machine to point **R**
N250 G40 X2.0 Y1.125	Cutter compensation off, point **S**
N260 G1 Z.1	Feed out of part
N265 G28 Z.1	Z Home point
N270 G0 X-4.0 Y0 Z5.0 M09	Move to tool change position
N280 M30	End of program
%	Rewind/stop code

TURNING CENTER PROJECTS

Turning and chucking centers are basically lathes equipped with computer numerical control (CNC) to guide the cutting tools and machine slides to perform various machining operations automatically. These machines are capable of greater precision and higher production rates than the standard lathe.

The projects in this section are designed for two types of machines:

1. The CNC bench-top teaching lathe that may have an EIA RS 274D or a Fanuc-compatible control.

2. The CNC standard turning center with a Fanuc-compatible control, commonly found in industry.

The basic programming for CNC lathes or turning centers is similar; however, some codes and minor programming procedures vary with the control found on each machine.

Parallel Turning

Parallel turning is one of the most common operations performed on turning centers. This project is programmed on a CNC bench-top teaching machine with a Fanuc compatible control. Some bench-top teaching machines have the cutting tool in front of the workpiece, and as a result, programming of X-moves and circular moves (**G02** and **G03**) differ from the standard turning centers. All dimensions are shown in decimal inches.

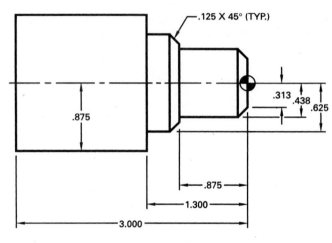

FIG. 14-1

OBJECTIVES

The parallel turning project should provide the following learning experiences:

1. Using the geometry offset for tool reference point.

2. Using the **X** code for setting depths of cuts (moving the tool toward or away from the work diameter).

3. Using the **Z** code for cutting to length.

4. The use of the **G00** code for the rapid positioning of the cutting tool.

Program Notes

1. Material: 2.00-in.-diameter brass (CS 500).

2. Diamond-shaped carbide cutting tool.

3. Use radius programming in the absolute mode.

4. All programming begins at the XZ zero at the centerline and right-hand end of the part.

5. Tool change position X1.250 Z.25.

6. Feed rate 10.00 in./min.

7. Spindle rotation counterclockwise (CCW).

The Program (Figure 14-1) (Fanuc Compatible Control, CNC Bench-Top Teaching Machine)

▲ CAUTION: Be sure this programming format suits your machine.

%	Rewind stop code/parity check
N010 G20 T0101 G90	Inch programming (Tool 1-geometry offset 01)
N030 S1.000 M03	Spindle on clockwise
N040 G00 X.800 Z.050	
N050 G01 X.800 Z-1.300 F.01	
N060 X.895 Z-1.300	
N070 G00 X.895 Z.050	
N080 X.750 Z.050	
N090 G01 X.750 Z-1.300	
N100 X.800 Z-1.300	
N110 G00 X.800 Z.050	
N120 X.625 Z.050	
N130 G01 X.625 Z-1.300	
N140 X.900 Z-1.300	
N150 G00 X.900 Z.050	
N160 X.500 Z.050	
N170 G01 X.500 Z-.875	
N180 X.625 Z-1.000	
N190 G00 X.625 Z.050	
N200 X.263 Z.050	
N210 G01 X.438 Z-.125	
N220 X.438 Z-.875	
N230 X.500 Z-.875 M09	
N240 G00 X1.250 Z.250	
N250 M30	
%	Rewind/stop code

Form Turning #1

Through the use of various command codes, it is possible to produce parallel diameters, chamfers, tapers, and curved forms on a part. This project uses a CNC bench-top teaching machine and the workpiece is held in a chuck for machining. When programming a part, always relate each machining operation to how it is done on a standard lathe; the machining procedures are very similar. All dimensions are shown in decimal inches.

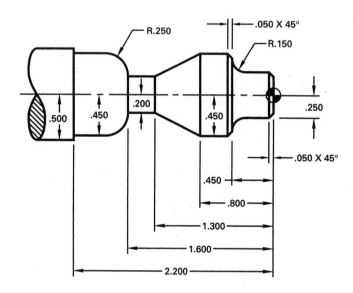

FIG. 15-1

OBJECTIVES

The form-turning #1 project should provide the following learning experiences:

1. Programming skills in linear and circular interpolation.

2. Using the **G02** and **G03** codes for producing curved sections (radii).

3. Using speed and feed codes.

Program Notes

1. Material: aluminum casting near-net size, approximately .025-in. oversize for machining (CS 500).

2. Diamond-shaped carbide cutting tool.

3. Radius programming in the absolute mode.

4. All programming starts at the XZ zero at the centerline and the right-hand end of the part.

5. Tool change position X1.25 Z.25.

6. Feed rate 10.000 in./min.

7. Spindle rotation counterclockwise (CCW).

The Program (Figure 15-1) (Fanuc Compatible Control, CNC Bench-Top Machine)

▲ **CAUTION: Be sure this programming format suits your machine.**	

%	Rewind stop code/parity check
N010 G20 T0101 G90	Inch programming/absolute (Tool 01, geometry offset 01)
N030 S 1000 M03	Spindle on clockwise (CW)
N040 G00 X0 Z.050	
N050 G01 X0 Z0 F.01	
N060 X.200 Z0	
N070 X.250 Z-.050	
N080 X.250 Z-.300	
N090 G03 X.400 Z-.450 1-.150 K0	
N100 G01 X.450 Z-.500	
N110 X.450 Z-.800	
N120 X.200 Z-1.300	
N130 X.200 Z-1.600	
N140 G02 X.450 Z-1.850 I0 K-.250	
N150 G01 X.450 Z-2.200	
N160 X.510 Z-2.200 M09	
N170 G00 X1.250 Z.500	
N180 M30	
%	Rewind/stop code

Form Turning #2

Most standard turning center controls have modal function capabilities, that is, they stay in effect in the program until they are replaced by another function code. Therefore, if a series of diameters (straight-line movements) is being machined using the G01 command code, it is not necessary to repeat this code in every program line. Standard turning centers have the cutting tool positioned at the back of the workpiece; therefore, the programming of X moves and circular moves differs from teaching-size machines.

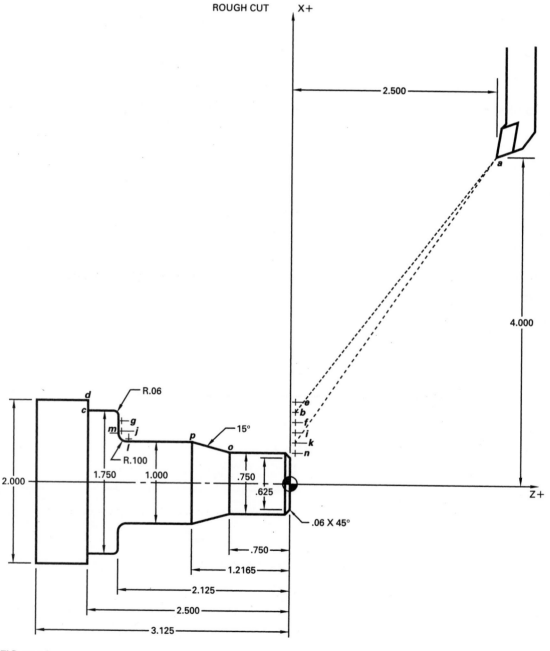

FIG. 16-1A

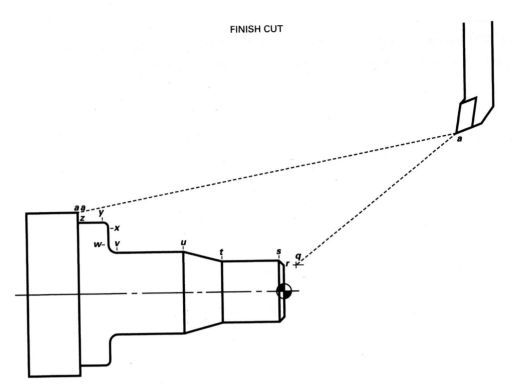

FINISH CUT

FIG. 16-1B

OBJECTIVES

The form-turning #2 project should provide the following learning experiences:

1. The use and value of modal codes.

2. Practice in linear interpolation along the X axis (tool infeed) and the Z axis (longitudinal movement) for machining parallel diameters, tapers, and chamfers.

3. Using **G01** and **G02** codes for machining circular forms clockwise and counterclockwise.

4. Using the geometry offset for tool position.

Program Notes

1. Diameter programming in the absolute mode (tool back of centerline).

2. All programming starts at the XY zero at the centerline and the right-hand end of the part.

3. Cutting tools: CNMG-432 for rough-turning, DNGG-432 for finish-turning.

4. The tool-change position is at X4.0 Z2.5.

5. Material AISI 1018 machine steel (CS 100).
   - (A) rough-turn at 400 sf/min.; feedrate .020-in./rev leaving .010-in. on shoulders and .020-in. on diameters for finishing.
   - (B) finish-turn at 800 sf/min.; feedrate .004-in./rev

The Program (Figure 16-1B)—(Standard CNC Turning Center—Fanuc Compatible Control)

%	Rewind stop code/parity check
N10 G20 T0100	Inch input data, activate geometry offset 01
N20 G50 S1500 M41	
G50 S1500	Maximum spindle speed 1500 r/min
M41	Low-speed range
N30 G00 T0101	
T0101	Activate offset number 01
N40 G96 S400 M03	
G96	Constant surface speed
S400	Speed 400 sf/min
M03	Spindle clockwise (CW)
N50 G00 X1.770 Z.1 M08	
G00	Rapid move **a-b**
M08	Coolant on
N60 G01 Z-2.490 F.020	
G01	Linear move **b-c**
F.020	Feed rate .020-in./rev
N70 X2.050	Linear move **c-d**
N80 G00 Z.1	Rapid move **d-e**
N90 X1.520	Rapid move **e-f**
N100 G01 Z-2.115	Linear move **f-g**
N110 X1.770	Linear move **g-h**
N120 G00 Z.1	Rapid move **h-b**
N130 X1.270	Rapid move **b-i**
N140 G01 Z-2.115	Linear move **i-j**
N150 X1.520	Linear move **j-g**
N160 G00 Z.1	Rapid move **g-f**
N170 X1.020	Rapid move **f-k**
N180 G01 Z-2.015	Linear move **k-l**
N190 X1.220 Z-2.115	Linear move **l-m**
N200 X1.270	Linear move **m-j**
N210 G00 Z.1	Rapid move **j-i**
N220 X.770	Rapid move **i-n**
N230 G01 Z-.740	Linear move **n-o**
N240 X1.020 Z-1.220	Linear move **o-p** (taper)
N250 G00 Z.1	Rapid move **s-k**
N260 G00 X4.0 Z2.5 M05	Rapid move **k-a**
M05	Spindle stop
N270 T0100	Activate offset number 00 (00 cancelled tool offset)

The Program—(Standard CNC Turning Center—Fanuc Compatible Control)

▲ CAUTION: Be sure this programming format suits your machine.

Finish Cut

N280 G50 S2000 M42 T0200	Activate geometry offset 02
N290 T0202 M06	
N300 G96 S700 M03	
S700	Speed 700 sf/min.
N310 G00 X.625 Z.2 T0202	Rapid to start of finish cut **a-q**
	Activate offset number 02
N320 G01 G42 Z0 F.004	
G42	Tool-nose radius compensation on during linear move **q-r**
N330 X.750 Z-.0625 F.004	Linear move **r-s** (chamfer)
N340 Z-.750	Linear move **s-t**
N350 X1.0 Z-1.2165	Linear move **t-u** (taper)
N360 Z-2.025	Linear move **u-v**
N370 G02 X1.200 Z-2.125 R.1	
G02	Circular interpolation (CW), **v-w**
R.1	.100-in. radius
N380 G01 X1.630	Linear move **w-x**
N390 G03 X1.750 Z-2.185 R.060	Circular interpolation counterclockwise **x-y**
N400 G01 Z-2.5	Linear move **y-z**
N410 X2.1 M09	Linear move **z-aa,** coolant off
N420 G00 G40 X4.0 Z2.5 M05	
G40	Cancel tool-nose radius compensation during rapid move **aa-a**
N430 T0200	Activate offset number 00 (00 cancelled tool offset 02)
N440 M30	End of program
%	Rewind/stop code

APPENDIX

TABLE 1   Decimal Inch, Fractional Inch, and Millimeter Equivalents

Decimal inch	Fractional inch	Millimeter	Decimal inch	Fractional inch	Millimeter
.015625	1/64	0.397	.515625	33/64	13.097
.03125	1/32	0.794	.53125	17/32	13.494
.046875	3/64	1.191	.546875	35/64	13.891
.0625	1/16	1.588	.5625	9/16	14.288
.078125	5/64	1.984	.578125	37/64	14.684
.09375	3/32	2.381	.59375	19/32	15.081
.109375	7/64	2.778	.609375	39/64	15.478
.125	1/8	3.175	.625	5/8	15.875
.140625	9/64	3.572	.640625	41/64	16.272
.15625	5/32	3.969	.65625	21/32	16.669
.171875	11/64	4.366	.671875	43/64	17.066
.1875	3/16	4.762	.6875	11/16	17.462
.203125	13/64	5.159	.703125	45/64	17.859
.21875	7/32	5.556	.71875	23/32	18.256
.234375	15/64	5.953	.734375	47/64	18.653
.25	1/4	6.35	.75	3/4	19.05
.265625	17/64	6.747	.765625	49/64	19.447
.28125	9/32	7.144	.78125	25/32	19.844
.296875	19/64	7.541	.796875	51/64	20.241
.3125	5/16	7.938	.8125	13/16	20.638
.328125	21/64	8.334	.828125	53/64	21.034
.34375	11/32	8.731	.84375	27/32	21.431
.359375	23/64	9.128	.859375	55/64	21.828
.375	3/8	9.525	.875	7/8	22.225
.390625	25/64	9.922	.890625	57/64	22.622
.40625	13/32	10.319	.90625	29/32	23.019
.421875	27/64	10.716	.921875	59/64	23.416
.4375	7/16	11.112	.9375	15/16	23.812
.453125	29/64	11.509	.953125	61/64	24.209
.46875	15/32	11.906	.96875	31/32	24.606
.484375	31/64	12.303	.984375	63/64	25.003
.5	1/2	12.7	1	1	25.4

TABLE 2 Conversions

Conversion of Inches to Millimeters						Conversion of Millimeters to Inches					
Inches	Milli-meters	Inches	Milli-meters	Inches	Milli-meters	Milli-meters	Inches	Milli-meters	Inches	Milli-meters	Inches
.001	0.025	.290	7.37	.660	16.76	0.01	.0004	0.35	.0138	0.68	.0268
.002	0.051	.300	7.62	.670	17.02	0.02	.0008	0.36	.0142	0.69	.0272
.003	0.076	.310	7.87	.680	17.27	0.03	.0012	0.37	.0146	0.7	.0276
.004	0.102	.320	8.13	.690	17.53	0.04	.0016	0.38	.015	0.71	.028
.005	0.127	.330	8.38	.700	17.78	0.05	.002	0.39	.0154	0.72	.0283
.006	0.152	.340	8.64	.710	18.03	0.06	.0024	0.4	.0157	0.73	.0287
.007	0.178	.350	8.89	.720	18.29	0.07	.0028	0.41	.0161	0.74	.0291
.008	0.203	.360	9.14	.730	18.54	0.08	.0031	0.42	.0165	0.75	.0295
.009	0.229	.370	9.4	.740	18.8	0.09	.0035	0.43	.0169	0.76	.0299
.010	0.254	.380	9.65	.750	19.05	0.1	.0039	0.44	.0173	0.77	.0303
.020	0.508	.390	9.91	.760	19.3	0.11	.0043	0.45	.0177	0.78	.0307
.030	0.762	.400	10.16	.770	19.56	0.12	.0047	0.46	.0181	0.79	.0311
.040	1.016	.410	10.41	.780	19.81	0.13	.0051	0.47	.0185	0.8	.0315
.050	1.27	.420	10.67	.790	20.07	0.14	.0055	0.48	.0189	0.81	.0319
.060	1.524	.430	10.92	.800	20.32	0.15	.0059	0.49	.0193	0.82	.0323
.070	1.778	.440	11.18	.810	20.57	0.16	.0063	0.5	.0197	0.83	.0327
.080	2.032	.450	11.43	.820	20.83	0.17	.0067	0.51	.0201	0.84	.0331
.090	2.286	.460	11.68	.830	21.08	0.18	.0071	0.52	.0205	0.85	.0335
.100	2.54	.470	11.94	.840	21.34	0.19	.0075	0.53	.0209	0.86	.0339
.110	2.794	.480	12.19	.850	21.59	0.2	.0079	0.54	.0213	0.87	.0343
.120	3.048	.490	12.45	.860	21.84	0.21	.0083	0.55	.0217	0.88	.0346
.130	3.302	.500	12.7	.870	22.1	0.22	.0087	0.56	.022	0.89	.035
.140	3.56	.510	12.95	.880	22.35	0.23	.0091	0.57	.0224	0.9	.0354
.150	3.81	.520	13.21	.890	22.61	0.24	.0094	0.58	.0228	0.91	.0358
.160	4.06	.530	13.46	.900	22.86	0.25	.0098	0.59	.0232	0.92	.0362
.170	4.32	.540	13.72	.910	23.11	0.26	.0102	0.6	.0236	0.93	.0366
.180	4.57	.550	13.97	.920	23.37	0.27	.0106	0.61	.024	0.94	.037
.190	4.83	.560	14.22	.930	23.62	0.28	.011	0.62	.0244	0.95	.0374
.200	5.08	.570	14.48	.940	23.88	0.29	.0114	0.63	.0248	0.96	.0378
.210	5.33	.580	14.73	.950	24.13	0.3	.0118	0.64	.0252	0.97	.0382
.220	5.59	.590	14.99	.960	24.38	0.31	.0122	0.65	.0256	0.98	.0386
.230	5.84	.600	15.24	.970	24.64	0.32	.0126	0.66	.026	0.99	.039
.240	6.10	.610	15.49	.980	24.89	0.33	.013	0.67	.0264	1	.0394
.250	6.35	.620	15.75	.990	25.15	0.34	.0134				
.260	6.60	.630	16	1	25.4						
.270	6.86	.640	16.26								
.280	7.11	.650	16.51								

TABLE 3   Letter Drill Sizes

Letters	in.	mm	Letters	in.	mm	Letters	in.	mm	Letters	in.	mm
A	.234	5.94	H	.266	6.76	N	.302	7.67	T	.358	9.09
B	.238	6.05	I	.272	6.91	O	.316	8.03	U	.368	9.35
C	.242	6.15	J	.277	7.04	P	.323	8.2	V	.377	9.58
D	.246	6.25	K	.281	7.14	Q	.332	8.43	W	.386	9.8
E	.250	6.35	L	.290	7.37	R	.339	8.61	X	.397	10.08
F	.257	6.53	M	.295	7.49	S	.348	8.84	Y	.404	10.26
G	.261	6.63							Z	.413	10.49

TABLE 4   Number Drill Sizes

Numbers	Inches	Millimeters	Numbers	Inches	Millimeters	Numbers	Inches	Millimeters
1	.228	5.79	34	.111	2.82	66	.033	0.84
2	.221	5.61	35	.110	2.79	67	.032	0.81
3	.213	5.41	36	.1065	2.71	68	.031	0.79
4	.209	5.31	37	.104	2.64	69	.0292	0.74
5	.2055	5.22	38	.1015	2.58	70	.028	0.71
6	.204	5.18	39	.0995	2.53	71	.026	0.66
7	.201	5.1	40	.098	2.49	72	.025	0.64
8	.199	5.05	41	.096	2.44	73	.024	0.61
9	.196	4.98	42	.0935	2.37	74	.0225	0.57
10	.1935	4.91	43	.089	2.26	75	.021	0.53
11	.191	4.85	44	.086	2.18	76	.020	0.51
12	.189	4.8	45	.082	2.08	77	.018	0.46
13	.185	4.7	46	.081	2.06	78	.016	0.41
14	.182	4.62	47	.0785	1.99	79	.0145	0.37
15	.180	4.57	48	.076	1.93	80	.0135	0.34
16	.177	4.5	49	.073	1.85	81	.013	0.33
17	.173	4.39	50	.070	1.78	82	.0125	0.32
18	.1695	4.31	51	.067	1.7	83	.012	0.3
19	.166	4.22	52	.0635	1.61	84	.0115	0.29
20	.161	4.09	53	.0595	1.51	85	.011	0.28
21	.159	4.04	54	.055	1.4	86	.0105	0.27
22	.157	3.99	55	.052	1.32	87	.010	0.25
23	.154	3.91	56	.0465	1.18	88	.0095	0.24
24	.152	3.86	57	.043	1.09	89	.0091	0.23
25	.1495	3.8	58	.042	1.07	90	.0087	0.22
26	.147	3.73	59	.041	1.04	91	.0083	0.21
27	.144	3.66	60	.040	1.02	92	.0079	0.2
28	.1405	3.57	61	.039	0.99	93	.0075	0.19
29	.136	3.45	62	.038	0.97	94	.0071	0.18
30	.1285	3.26	63	.037	0.94	95	.0067	0.17
31	.120	3.05	64	.036	0.91	96	.0063	0.16
32	.116	2.95	65	.035	0.89	97	.0059	0.15
33	.113	2.87						

TABLE 5 Commercial Tap Drill Sizes (75% of Thread Depth) American National and Unified Form Thread

NC National Course*			NF National Fine*		
Tap size	Threads per inch	Tap drill size	Tap size	Threads per inch	Tap drill size
# 5	40	#38	# 5	44	#37
# 6	32	#36	# 6	40	#33
# 8	32	#29	# 8	36	#29
#10	24	#25	#10	32	#21
#12	24	#16	#12	28	#14
$1/4$	20	# 7	$1/4$	28	# 3
$5/16$	18	F	$5/16$	24	I
$3/8$	16	$5/16$	$3/8$	24	Q
$7/16$	14	U	$7/16$	20	$25/64$
$1/2$	13	$27/64$	$1/2$	20	$29/64$
$9/16$	12	$31/64$	$9/16$	18	$33/64$
$5/8$	11	$17/32$	$5/8$	18	$37/64$
$3/4$	10	$21/32$	$3/4$	16	$11/16$
$7/8$	9	$49/64$	$7/8$	14	$13/16$
1	8	$7/8$	1	14	$15/16$
$1\,1/8$	7	$63/64$	$1\,1/8$	12	$1\,3/64$
$1\,1/4$	7	$1\,7/64$	$1\,1/4$	12	$1\,11/64$
$1\,3/8$	6	$1\,7/32$	$1\,3/8$	12	$1\,19/64$
$1\,1/2$	6	$1\,11/32$	$1\,1/2$	12	$1\,27/64$
$1\,3/4$	5	$1\,9/16$			
2	$4\,1/2$	$1\,25/32$			
NPT National Pipe Thread					
$1/8$	27	$11/32$	1	$11\,1/2$	$1\,5/32$
$1/4$	18	$7/16$	$1\,1/4$	$11\,1/4$	$1\,1/2$
$3/8$	18	$19/32$	$1\,1/2$	$11\,1/2$	$1\,23/32$
$1/2$	14	$23/32$	2	$11\,1/2$	$2\,3/16$
$3/4$	14	$15/16$	$2\,1/2$	8	$2\,5/8$

*The major diameter of an NC or NF number size tap or screw = (N × .013) + .060.
Example: The major diameter of a #5 tap equals (5 × .013) + .060 = .125 diameter.

TABLE 6   Commonly Used Formulas

Code

c.p.t = Chip per tooth	N = Number of threads per inch	T.D.S. = Tap drill size
CS = Cutting speed	= Number of strokes per minute	T.L. = Taper length
D = Large diameter	= Number of teeth in cutter	tpf = Taper per foot
d = Small diameter	O.L. = Overall length of work	tp/mm = Taper per millimeter
G.L. = Guide bar length	P = Pitch	T.O. = Tailstock offset

Inch	Metric
$\text{T.D.S.} = D - \left(\dfrac{1}{N}\right)$	$\text{T.D.S.} = D - P$
$\text{r/min} = \dfrac{\text{CS (ft)} \times 4}{D \text{ (in.)}}$	$\text{r/min} = \dfrac{\text{CS (m)} \times 320}{D \text{ (mm)}}$
$\text{tpf} = \dfrac{(D - d) \times 12}{\text{T.L.}}$	$\text{tp/mm} = \dfrac{(D - d)}{\text{T.L.}}$
$\text{T.O.} = \dfrac{\text{tpf} \times \text{O.L.}}{24}$	$\text{T.O.} = \dfrac{\text{tp/mm} \times \text{O.L.}}{2}$
$\text{Guide bar setover} = \dfrac{(D - d) \times 12}{\text{T.L.}}$	$\text{Guide bar setover} = \dfrac{(D - d)}{2} \times \dfrac{\text{G.L.}}{\text{T.L.}}$
Milling feed (in./min) = $N \times$ c.p.t $\times$ r/min	Milling feed (mm/min) = $N \times$ c.p.t $\times$ r/min

Source: Krar et al., *Machine Tool Operations*, p. 391.

TABLE 7   Formula Shortcuts

For the correct formula, block out (cover) the unknown; the remainder is the formula. In each diagram the horizontal line is the division line; the vertical line is the multiplication line.

Code: A = Area	L = Length	b = Base
C = Circumference	R = Radius	h = Height
CS = Cutting speed	r/min = Revolutions/minute	m = Meters
D = Diameter		mm = Millimeters

Circle

$C = \pi \times D$

$D = \dfrac{C}{\pi}$

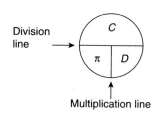

Four-Element Formulas
1. Block out unknown.
2. Cross-multiply diagonally opposite elements.
3. Divide by remaining element.

Triangles

$A = \dfrac{b \times h}{2}$

$b = \dfrac{A \times 2}{h}$

$h = \dfrac{A \times 2}{b}$

Area

Squares and rectangles	Circles
$A = L \times W$ $L = \dfrac{A}{W}$ $W = \dfrac{A}{L}$	$A = \pi \times R^2$ $R^2 = \dfrac{A}{\pi}$

Revolutions per Minute (r/min)
(Lathe, drill, mill, grinder)

Inch	Metric
$\text{r/min} = \dfrac{\text{CS (ft)} \times 4}{D \text{ (in.)}}$ $\text{CS} = \dfrac{\text{r/min} \times D}{4}$ $D = \dfrac{\text{CS} \times 4}{\text{r/min}}$	$\text{r/min} = \dfrac{\text{CS (m)} \times 320}{D \text{ (mm)}}$ $\text{CS} = \dfrac{\text{r/min} \times D}{320}$ $D = \dfrac{\text{CS} \times 320}{\text{r/min}}$

Source: Krar et al., *Machine Tool Operations*, p. 390.

TABLE 8   Three-Wire Thread Measurement (60° Metric Thread)

M = PD + C   PD = M − C

M = Measurement over wires
PD = Pitch diameter
 C = Constant

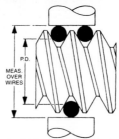

Pitch		Best Wire Size		Constant	
Inches	mm	Inches	mm	Inches	mm
.00787	0.2	.00455	0.1155	.00682	0.1732
.00886	0.225	.00511	0.1299	.00767	0.1949
.00934	0.25	.00568	0.1443	.00852	0.2165
.01181	0.3	.00682	0.1732	.01023	0.2598
.01378	0.35	.00796	0.2021	.01193	0.3031
.01575	0.4	.00909	0.2309	.01364	0.3464
.01772	0.45	.01023	0.2598	.01534	0.3897
.01969	0.5	.01137	0.2887	.01705	0.433
.02362	0.6	.01364	0.3464	.02046	0.5196
.02756	0.7	.01591	0.4041	.02387	0.6062
.02953	0.75	.01705	0.433	.02557	0.6495
.0315	0.8	.01818	0.4619	.02728	0.6928
.03543	0.9	.02046	0.5196	.03069	0.7794
.03937	1	.02273	0.5774	.0341	0.866
.04921	1.25	.02841	0.7217	.04262	1.0825
.05906	1.5	.0341	0.866	.05114	1.299
.0689	1.75	.03978	1.0104	.05967	1.5155
.07874	2	.04546	1.1547	.06819	1.7321
.09843	2.5	.05683	1.4434	.08524	2.1651
.11811	3	.06819	1.7321	.10229	2.5981
.1378	3.5	.07956	2.0207	.11933	3.0311
.15748	4	.09092	2.3094	.13638	3.4641
.17717	4.5	.10229	2.5981	.15343	3.8971
.19685	5	.11365	2.8868	.17048	4.3301
.21654	5.5	.12502	3.1754	.18753	4.7631
.23622	6	.13638	3.4641	.20457	5.1962
.27559	7	.15911	4.0415	.23867	6.0622
.31496	8	.18184	4.6188	.27276	6.9282
.35433	9	.20457	5.1962	.30686	7.7942
.3937	10	.2273	5.7735	.34095	8.6603

TABLE 9　Tapers and Angles

Taper per Foot	Include Angle		With Center Line		Taper per Inch	Taper per Inch from Center Line
	Degree	Minute	Degree	Minute		
1/8	0	36	0	18	.010416	.005208
3/16	0	54	0	27	.015625	.007812
1/4	1	12	0	36	.020833	.010416
5/16	1	30	0	45	.026042	.013021
3/8	1	47	0	53	.03125	.015625
7/16	2	05	1	02	.036458	.018229
1/2	2	23	1	11	.041667	.020833
9/16	2	42	1	21	.046875	.023438
5/8	3	00	1	30	.052084	.026042
11/16	3	18	1	39	.057292	.028646
3/4	3	35	1	48	.0625	.03125
13/16	3	52	1	56	.067708	.033854
7/8	4	12	2	06	.072917	.036458
15/16	4	28	2	14	.078125	.039063
1	4	45	2	23	.08333	.041667
1 1/4	5	58	2	59	.104166	.052083
1 1/2	7	08	3	34	.125	.0625
1 3/4	8	20	4	10	.145833	.072917
2	9	32	4	46	.166666	.083333
2 1/2	11	54	5	57	.208333	.104166
3	14	16	7	08	.250	.125
3 1/2	16	36	8	18	.291666	.145833
4	18	56	9	28	.333333	.166666
4 1/2	21	14	10	37	.375	.1875
5	23	32	11	46	.416666	.208333
6	28	04	14	02	.500	.250

Courtesy Morse Twist Drill & Machine Co.

TABLE 10 Solutions for Right-Angle Triangles

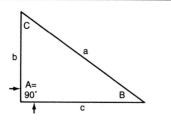

$\text{Sine}\angle = \dfrac{\text{Side opposite}}{\text{Hypotenuse}}$	$\text{Cosecant}\angle = \dfrac{\text{Hypotenuse}}{\text{Side opposite}}$
$\text{Cosine}\angle = \dfrac{\text{Side adjacent}}{\text{Hypotenuse}}$	$\text{Secant}\angle = \dfrac{\text{Hypotenuse}}{\text{Side adjacent}}$
$\text{Tangent}\angle = \dfrac{\text{Side opposite}}{\text{Side adjacent}}$	$\text{Cotangent}\angle = \dfrac{\text{Side adjacent}}{\text{Side opposite}}$

Knowing	Formulas to Find	
Sides a and b	$c = \sqrt{a^2 - b^2}$	$\sin B = \dfrac{b}{a}$
Side a and angle B	$b = a \times \sin B$	$c = a \times \cos B$
Sides a and c	$b = \sqrt{a^2 - c^2}$	$\sin C = \dfrac{c}{a}$
Side a and angle C	$b = a \times \cos C$	$c = a \times \sin C$
Sides b and c	$a = \sqrt{b^2 + c^2}$	$\tan B = \dfrac{b}{c}$
Side b and angle B	$a = \dfrac{b}{\sin B}$	$c = b \times \cot B$
Side b and angle C	$a = \dfrac{b}{\cos C}$	$c = b \times \tan C$
Side c and angle B	$a = \dfrac{c}{\cos B}$	$b = c \times \tan B$
Side c and angle C	$a = \dfrac{c}{\sin C}$	$b = c \times \cot C$